THE MYSTERY OF TIME

THE MYSTERY
OF TIME

HUMANITY'S QUEST
FOR ORDER AND MEASURE

JOHN LANGONE

NATIONAL
GEOGRAPHIC

WASHINGTON, D.C.

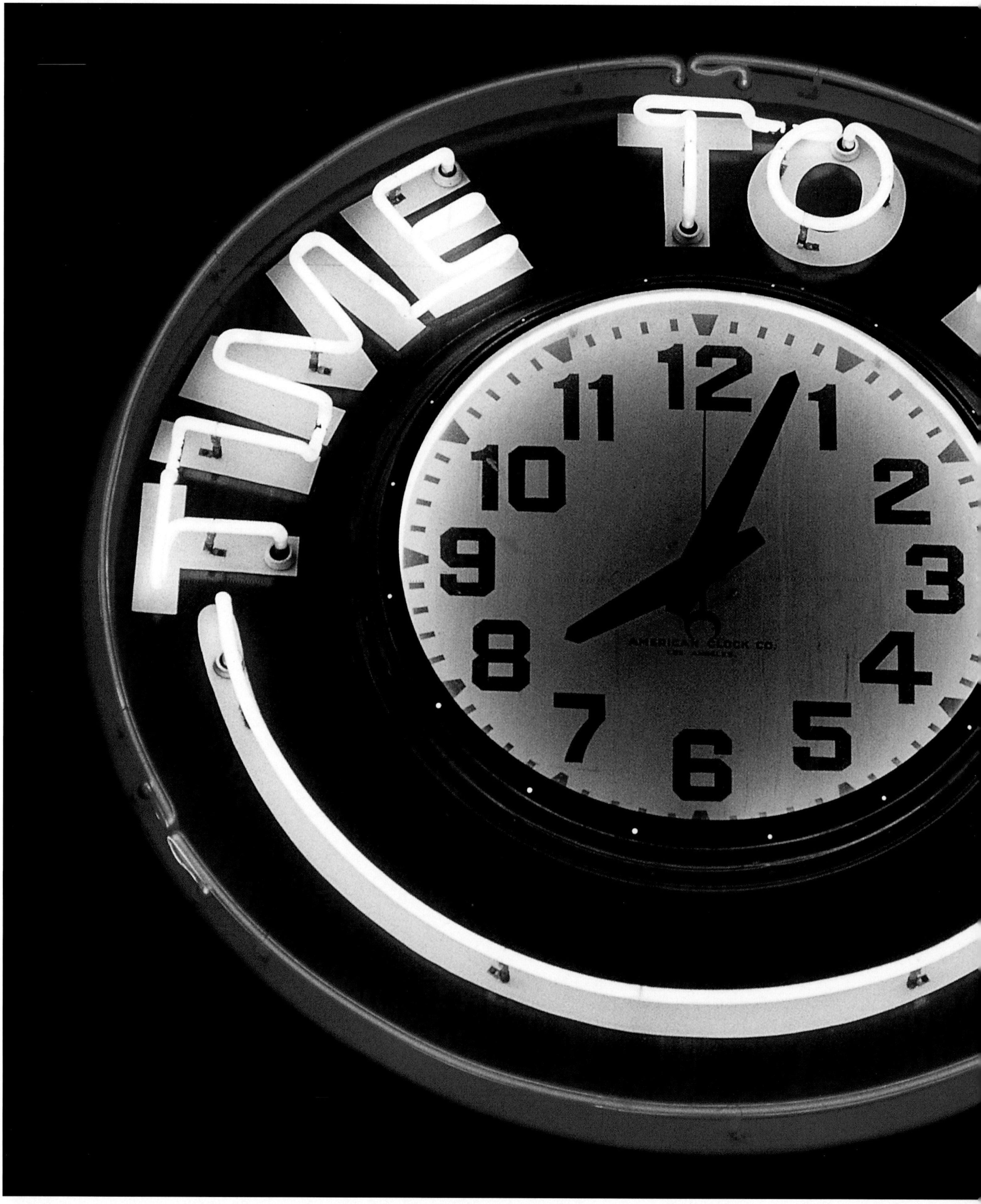
TIME TO
AMERICAN CLOCK CO.

BRENTON

CONTENTS

NEON TIMEPIECE BLAZES ATOP A CAFÉ ON LOS ANGELES'S SANTA MONICA BOULEVARD.

PAGE 1: **TIME'S PASSAGE** IS RECORDED IN THE GROWTH RINGS OF A CHAMBERED NAUTILUS SHELL.

PAGES 2-3: **AGELESS BEAUTY** OF A GERMAN FOREST AWES A NEW GENERATION OF CHILDREN.

TIME: THE ELUSIVE CONCEPT

"What, then, is time?
If no one asks me, I know.
If I wish to explain it to one that asketh,
I know not."

— ST. AUGUSTINE OF HIPPO

ST. AUGUSTINE, as depicted by the Venetian painter Vittore Carpaccio (opposite), is apparently engrossed in his thoughts. In his monumental *City of God,* the fourth-century monk addressed the meaning of the past and the secrets of the future. But centuries later, his questions concerning the nature of time remain unanswered. In a drawing from 1500 (above), a bird symbolizes air, an element as elusive as time itself to early philosophers.

WHEN ST. AUGUSTINE OF HIPPO, the fourth-century church father and philosopher, pondered the nature of time, even he, who had much to say about many things, expressed puzzlement. "Who," he asked, "can even in thought comprehend it, so as to utter a word about it?… My soul is on fire to know this most intricate enigma."

Those words resonate today. Of all the scientific intangibles that shape our lives, time is arguably the most elusive—and the most powerful. As formless as space and being, those other unseen realms of abstraction on which we are helplessly dependent, it nonetheless affects all material things—along with our immaterial minds. Its power is awesome. Without it we could barely measure change, for most things that change on this Earth and in the universe happen in time and are governed by it. Stealthy, imperceptible, time makes its presence known by transforming our sense of it into sensation. For though we cannot see, touch, or hear time, we observe the regularity of what appears to be its passage in our seasons, in the orchestrated shift from dawn to dusk to dark, and in the aging of our bodies. We feel its pulsing beat in our hearts and hear its silence released in the precise ticking of a clock.

But the question remains: What is it? While a lot of us believe in "true" time, we aren't certain it exists, and, if it does, we don't know whether it is psychological or physical. The 11th-century Islamic scientist and philosopher Avicenna argued that time existed only in the mind, based on our memories and expectations. Isaac Newton, the father of modern physics, said it was a substance. In this book we look at the many ways that time makes itself manifest in nature and at the ways that humans attempt to measure it, use it, interpret it, escape it, manipulate it, ruminate on it, or travel through it.

WITHOUT TIME AND THE TOOLS we use to measure it—clocks, calendars, speedometers, chronometers, vibrating atoms, and a host of other artifacts that mark it—life, as we live it today, would be severely unregulated to say the least. We would not know what time to eat and what time to sleep, what time to work and what time to play. Scientific observation would be impossible, since that often depends on exactly when an event occurs within a given time frame and on how long it lasts. As Princeton physicist John Wheeler once observed, time is nature's way of keeping everything from happening at once.

Time, thus, brings order to chaos, and structure, meaning, and continuity to our lives. Its involvement is indirect, for, as far as we know, time doesn't really do anything, though we like to say it does. The late Harvard psychologist B. F. Skinner probably put it best when he observed, "Wounds heal in time, things are constructed in time, things disappear in time and are destroyed in time, but this is not what time does." Still, when talking about time, it's difficult not to use words that give it substance. So, for our purposes, let's say that it establishes intervals and periods. It arranges our lives. It alerts us to the "now" and the "then." And, with apologies to B. F. Skinner, let's also insist that time does heal physical wounds and psychic grief. The 19th-century British politician Benjamin Disraeli dubbed time the "great physician." It forces us to learn patience and gives us a past and a history to draw on, a present to savor, and a future with potential and promise. Time can also be maddeningly miserly, as when it doles out pleasure, often permitting only happy moments and not happy years. Merciless, too, it lets us know in no uncertain terms that all things pass away when their allotted time runs out, that nothing—perhaps not even time itself—is eternal.

Time is on our lips almost every day of our lives. We say that we have it on our hands, or that we are wasting it; we double it, value it, mark it, share it, limit it, flex it, find it, save it, kill it, lose it, take it, have it, bide it, run out of it, and keep it perfect. We are sometimes ahead of it, sometimes

behind. Things are time-tested, timeworn, and time-honored. There is time in bombs, zones, capsules, lines, sheets, exposures, charts, and calendars.

BUT BACK TO ST. AUGUSTINE'S question: What, exactly, is time? Is it physical or in our minds? Is all time the same, no matter who is experiencing it? How did it begin? Where does it go? Does it really move, as so many people believe it does? If so, does it move in only one direction—in a straight line, forward? Or does it go around, as in a clock, or in cycles? Is it an addition to the length, breadth, and width known as the fourth dimension, an unrectangular coordinate outside the range of ordinary experience? What is it doing now, as you read this? What is its speed? A second per second? Can we slow it down, reverse it, stop it? Does the present have any duration at all? Does the future exist now? Will time ever end? Is dream time real time? Finally, we might ask if time can exist without events or change, whether it really exists at all, and whether it is no more than just what appears on a clock's face.

The questions are neither frivolous inquiries nor mindless wordplay. Physicists, theologians, and philosophers have wrestled with many of them for centuries, but, while lofty surmises abound, definitive answers, if there are any at all, will only be found—dare we say it?—in their own time. The major obstacle is, of course, the lack of anything tangible about time. Like nothingness, it has no separable elements, no landmark objects to give it meaning. We cannot stain time to make it visible, nor catch it in a bottle, nor subject it to spectroscopic analysis. Its alleged passage can only be recorded by the tools of timekeeping—and sensed by us humans with some still obscure inner mechanism—and by terms of our own making.

Even a scientist of the stature of the late Richard P. Feynman, winner of the 1965 Nobel Prize in physics, was at a loss to define time. In his celebrated lectures on physics at Caltech, he described how Albert Einstein changed the prevailing view of space and time as separate things into a combined space-time. Yet, like St. Augustine, when Feynman pondered what time was, his observation was similar to that made centuries ago by the theologian. "We physicists work with it every day," he said, "but don't ask me what it is. It's just too difficult to think about."

But think about it we must, for time, though it is murkier territory than even space, is capable of being grasped in two ways by our minds. There is, some suggest, the subjective, inner world of time— soul time, it is sometimes called—whose passage and duration are somehow felt. It is personal and private, the sort of time that flies when we are having fun or slows when we are waiting for the pot to boil or pain and anguish to end. As events play out and changes occur, as our thoughts pass and time seems to lapse, we sense a loss of something that was here and has gone by.

Then there is the objective, external world of "public" or "clock" time, the measurable, physical time we live by. It is a signaling system, a substitution for, and an imagined response to, what some have called "pure" time; this line of reasoning suggests that we do not truly feel the passage of a fictitious clock time that ticks away our hours. We are simply going along with a spatial notion—the position of the clock's hands. The seconds that tick by are not the sounds of time passing. Referring to a clock, then, merely tells us what time it is—not what time is. As Aristotle put it, "Time obviously is what is counted, not that with which we count." Longfellow, too, noted the difference between the two perceptions of time. "What is time?" he asked. "The shadow on the dial, the striking of the clock, the running of the sand, day and night, summer and winter, months, years, centuries—these are but the arbitrary and outward signs, the measure of time, not time itself. Time is the life of the soul."

More subtle and experiential, the idea of time sense, the mind's perception of time, has a philosophical and psychological basis that relegates time to a compartment of our consciousness. "Measure, time, and number are nothing but modes of thought, or rather of imagination," said the Dutch philosopher Benedict Spinoza.

Since our inner sense of time's passage quits when our consciousness does, one might be tempted to argue that time's existence is, indeed, dependent on someone thinking about it. We make it exist. As Aristotle explained it, "Whether, if soul did not exist, time would exist or not, is a question that may be fairly asked. For if there cannot be someone to count, there cannot be anything that can be counted."

Two of the most notorious obstacles to unravelling the concept of time are the notions of duration and of time as an absolute flowing, a reality apart from the events that fill it, with a fixed, uniform rate in which all change may be measured.

First, duration. Duration can mean the continuance in time, time as a whole, or the time in which something exists or lasts. It has been defined as more than the momentary part of time and as eventless time. The 17th-century English philosopher John Locke said it was "fleeting extension," a characterization elaborated upon in more recent times by Henri-Louis Bergson. The French philosopher made duration a central part of his philosophy, buttressing the subjective conception of time we discussed earlier as the "something real" that is experienced within us. Duration, he believed, is intuitively grasped, a stream of consciousness in which past, present, and future flow into one another. It was also somehow connected to memory, since in

the memory, it is the past and not the future that survives. Duration is, thus, closely linked to time's passage, to flow, to what Plato referred to as "the moving image of eternity." It is," Plato said, "as if we were floating on a river, carried by the current past the manifold of events which is spread out timelessly on the bank." For Aristotle, time's passage was "the number of movements in respect of the before and after, and is continuous." The 17th-century French prelate St. Francis de Sales lifted it to another plane: "Let time flow by," he said, "with which we flow on to be transformed into the glory of the children of God."

Early references to a flow of time and its absolute nature, whether they derived from scientists or philosophers, would continue to come from both until Albert Einstein successfully challenged them in 1905 with his special theory of relativity. Rhetoric about time's flow would also contribute to a trove of "philosospeak" and contradictions that persist to this day.

Consider the discussion of the nature and behavior of the present, also called the "now," which refers to a moment in time that changes. Not a word with any counterpart in scientific laws, now, true to its name, shows up not now and then, but now and again in the philosophical literature. If, as one timely argument swirling around it goes, we accept that the present flows to a future time, we are saying that this very moment of time, this now, will become some other moment and cease to be identical with itself. But, since every moment is a now when it occurs, the notion is deemed absurd. If time is composed of continuous instants, linearly ordered from past to future, and they are flowing along, when does the present instant cease to exist? One answer is "not at the present," because, while the instant exists, it exists. But it can't cease to exist at the next moment, because in a continuum there is no next moment; nor can it cease to exist in any subsequent moment because then it is already gone. But how can the present instant continuously exist? If that were the case, events that occurred centuries ago would be simultaneous with things that have happened today. Is it any wonder that 16th-century French essayist Montaigne said that to philosophize is to doubt?

Because time is so shadowy, it needs attributes to be conceived at all. So again we have had to give it some substance to fathom it; hence the allusions to time as a stream or river on its way from source to an emptying end. We also give it its other attributes of past, present, and future, and these become the convenient locations, neatly constructed way stations, if you will, where the events carried along in the flow can play out their assigned roles. Ruminations that dwell on time "slipping by" are logical—if events are seen as approaching us from the future, pausing momentarily in the present, and disappearing into the past. Birthdays are anticipated, they arrive from the future and are celebrated, and then they go by.

It should be obvious, then, why we see time as an absolute—something that flows perpetually in one direction, unrestrained, and at a uniform rate. But how can we be certain that time truly flows from past to future rather than from future to past?

DOES TIME REALLY FLOW at all, though? Or is its apparent passage merely a figment of our imaginations, a perception, a feature of subjective, psychological time? First of all, since neither philosophers nor physicists can really explain time, physical or psychological, any answer is often speculation based on speculation. Still, there are some valid reasons for discounting the popular idea of time's flow. One is, of course, the lack of any real way to actually prove it flows—how does one tell if something is moving if it can't be seen, heard, or felt, and, indeed, may not even exist as an entity at all?

Time's flow is on shaky ground for another reason: If it does flow, how fast does it actually travel? Let's return to the question of how one is supposed to tell when an instant ceases to be. If time is a linear continuum composed of instants, there'd be no next instant. That's because the instants would be packed too closely together to be associated with numbers. Measuring the motion of something requires an allusion to time in relation to space, and so we talk of light traveling at more than 186,000 miles per second, and a car speeding along at 65 miles an hour. But how does one measure the passage of something that is not only composed of an infinity of moments that modern technology can subdivide limitlessly but is the gauge itself? Where would you fix a point from which to begin, since time is not on any map? What unit of measurement would you use anyway? Since associating time with miles is meaningless, one cannot say time travels so many miles per nanosecond. A second per second may sound right, or a nanosecond per nanosecond, but those are examples of what philosophers describe as tautology, a round-robin kind of logic that turns on itself. "If time flows past us," an entry in the comprehensive *Encyclopedia of Philosophy* observes, "or if we advance through time, this would be motion with respect to a hyper-time (a sort of super-time).... Moreover, if passage is the essence of time, it is presumably the essence of hypertime too, which would lead us to postulate a hyper-hypertime, and so on ad infinitum."

"If you are of a naturalistic turn of mind," MacNeile Dixon explains, "you are plunged in still deeper distress. For you have to determine the position in space and time not only of material objects, but of your ideas and sentiments. How are they to be either located or measured?" So, scrap putting a

PASSAGE OF TIME,
as it moves inexorably
through human
civilization, is shown
in this fanciful triptych.
At the left, an
Alexandrian doctor
in the year A.D. 1
examines a youth;
neither has any inkling
that the lighthouse
in the background,
a monument to
invincibility, will eventually topple into the sea.
A thousand years later,
an Islamic scholar
(center) teaches two
children the Koran
in Córdoba, a city
conquered by the Moors
300 years before.
A thousand years
further on, in a vastly
changed world,
a woman sits in her
basement apartment
in New York City, and
connects to the outside
world via computer,
cell phone, television,
and radio.

speedometer on time, itself the ultimate measuring tool. If it does go from somewhere way out there, to here, to somewhere way back, then there isn't an instrument known to humankind that can detect such movement.

What about how events and objects change? Doesn't that prove that time not only exists but passes steadily? After all, we do seem to sense time passing only when something has happened, not happened, or changed. We all know that nothing that comes to us in this world stays. Ring a bell once, hear the note dissipate, and you'll be instantly aware of that fact. Things change, times change. Wars and World Series come and go. People, rocks, and houses wear away. We change our clothes, way of living, position, moods, facial expressions, diets, and minds. All in time. The impermanence that numbers our days, it seems, is part of life. Some 500 years before Christ, the pre-Socratic Greek philosopher, Heraclitus, argued, in fact, that there was no permanent reality but the reality of change, which characterized everything, and that the only possible real state was the transitional one of becoming.

We measure the evolution of a physical system in terms of time. That being the case, time can't exist independent of events; time is, thus, formed by events and the relations among them. Aristotle tied the concept of change directly to time when he argued: "Neither does time exist without change, for when the state of our own minds does not change at all, or we have not noticed it changing, we do not realize that time has elapsed." When we see a change in something tangible—erosion of a beach, say, or the graying of our hair—we are able to perceive a difference between before and now. However, while people, books, and automobile tires change noticeably over time under external and internal influences, we might say "it's time" for these changes to happen, but we don't usually speak

of them as becoming past. Instead, we say people have gotten old, that books have fallen apart, that tires have worn out. It is the same way that nature, operating under the rule of the Second Law of Thermodynamics, makes things go from order to disorder—not how time flows between a vague before into now.

Let's return to events, which are another matter. We commonly think of them as having, in turn, a past, present, and future, characteristics that inexplicably shift mechanically and neatly into successive notches as time passes. Events are happenings (social activities, say) to "continuants"—that can cause change in things. An object changes its properties during an event. One philosopher offers this example: "The event of the buttering of toast involves the toast's changing from having the property of being unbuttered to having the property of being buttered." But, unlike objects, the events themselves do not change any genuine properties. An event's change from past to present to future, say some philosophers, is only in relation to an observer at a particular time and is not real change, as in the change that occurs when perfume evaporates or a raft floats downstream. Thus, change in an event cannot be used to prove that time passes, since that would assume the impossibility of change itself changing.

Memories bear some mention here. They are recollections of things and events and have been cited by Bergson and others as evidence of passing time. But, like the idea of events changing, a memory-connection to time is fanciful. Bergson's view was challenged by English philosopher Bertrand Russell, who felt that Bergson confused the memory of the past event with the past event itself, or the thought with that which is thought about. In any case, we remember only past events, not future ones. As to the whereabouts of

GEOCENTRIC COSMOLOGY, espoused by the Alexandrian astronomer Ptolemy and embraced by early Christianity, envisioned Earth as the center of the universe. That the concept had the approval of divine hosts is evidenced by this German woodcut of 1493 (opposite, left), showing God and the heavenly court arrayed around the Earth. The belief held sway over Western thought until the 16th century, when Copernicus displaced Earth in favor of a sun-centered, or heliocentric, vision. A 17th-century Dutch engraving (opposite, right) pays homage to the Copernican universe, showing the Earth orbiting a stationary sun.

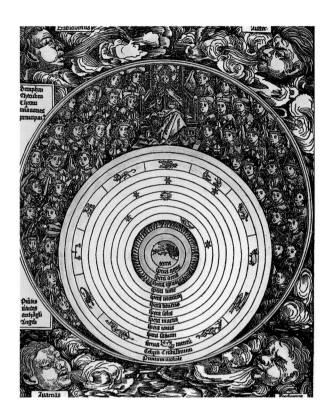

all those events we cannot recall… well, that is beyond the grasp of everyone on this Earth.

It seems logical to assume that if the events were caught in a time flow, they'd wash up somewhere in our consciousnesses. But nobody knows what memories are made of, beyond their being the product of chemical and other changes in our brain. Whenever there's a new experience in our lives, or we learn something, chemical traces are laid down, stored, and recalled out of our neurons. Thus, they are no less than a biochemical recording of what we have learned, seen, smelled, and heard in a given present, in a "now," an imprint of sights, sounds, and smells. Memories are not evidence of something being literally carried along in an abstraction like time. Perhaps it is best to just accept the simple observation Leonardo da Vinci made in the 15th century: "Good memory wherewith Nature has endowed

us causes everything long past to seem present." Memory, after all, may just be memory.

In the last analysis, one can safely say that the notion of a substantive time, as a property of the physical universe composed of some mysterious particles passing and slipping by us like a stream, is a metaphor and a myth, much of it based on confusion. Even if time did march on, it would unquestionably do so to its own inner drummer and without tramping feet or flags to herald its progress. The idea of flowing, palpable time is also a dodge, since we cannot find the right words to describe its "form," and since our human intelligence is not capable of fully understanding it, no more than we can understand how shapeless consciousness is present in our physical brains. Although philosophy has given us much that is provocative about time, it has provided little that is concrete, and it is doubtful

TIME'S GREAT THINKERS

PYTHAGORAS *(ca 580–500 B.C.)*
Numbers and regularity, two of time's central components, intrigued the ascetic and scientist from Samos, Greece. Pythagoras contemplated an objective world in which numerical proportions were repeated in many phenomena, according to a predictable scheme. Fundamentally mystical, Pythagoras's theories nonetheless included an awareness of the different orbital times of planets and the fact that the Earth travels regularly around the sun.

PTOLEMY *(Second Century A.D.)*
Astronomer, mathematician, and geographer, Ptolemy suffered a noble failure with his geocentric description of the universe, but his monumental mathematical compilation, *Almagest,* nonetheless reigned as the bible of mathematical astronomers until the 17th century. He also applied mathematics to the construction of sundials and devised a system of longitude and latitude.

GREGORY XIII *(1502–1585)*
Born Ugo Buoncompagni and elected pope in 1572, Gregory reformed the Julian calendar because its cumulative errors had skewed astronomical observations and displaced Christian feast days. To correct the errors, Gregory simply struck ten days from the old calendar in 1582 and revised the way in which leap years were calculated.

GALILEO GALILEI *(1564–1642)*
Promulgator of the laws governing the behavior of falling bodies, this Italian physicist/mathematician also discovered a pendulum's isochronism (the timing of its swing)—an observation that would lead to the most accurate clock regulator yet developed.

CHRISTIAAN HUYGENS *(1629–1695)*
A Dutch physicist, astronomer, and

Guided by Urania, the muse of astronomy, the Alexandrian astronomer Ptolemy uses a quadrant to measure the altitude of the moon.

mathematician, Huygens authored *Horologium Oscillatorium,* addressing problems involving the rotation of bodies, centrifugal force, and, more specific to time, pendulums. In 1656, he became the first to design a clock that used a pendulum to regulate its movements. Later, Huygens designed a balance wheel and spring assembly that is still used in some wristwatches today.

ISAAC NEWTON *(1642–1727)*
The genius who discovered the calculus and stated the basic laws of gravitation and motion, Newton would be proved wrong when he argued that time was absolute: that is, that the duration of time between two events would always be the same no matter who measured it. But Newton's work

with acceleration—a change in velocity with time—defined not only the relationship of motion to time but of both to force.

ALBERT EINSTEIN *(1879–1955)*
If anyone deserves the appellation "Father Time," it is the German-born, Swiss-American theoretical physicist whose name is also synonymous with relativity. His revolutionary theories made science think anew its understanding of time, space, mass, motion, and gravitation. His recognition that the speed of light is the same for all observers, no matter how they are moving, led to the realization that time is not, as Newton supposed, absolute. Rather, time is personal and relative.

ISIDOR I. RABI *(1898–1988)*
The Austrian-American chemist is perhaps best known for his Nobel Prize-winning work on the resonance method of measuring the magnetic properties of atoms and molecules. One practical application was the atomic clock, a device that uses the remarkable constancy of the electromagnetic radiation emitted by vibrating atoms as a way to "count" increments of time. Extremely accurate, atomic clocks serve as the ultimate standard for time telling throughout the world.

STEPHEN HAWKING *(1942–)*
Born on the anniversary of Galileo's death and holder of Newton's chair at Cambridge University, Hawking has given the world a new look at time, space, and the very shape of the universe. Hawking's theoretical universe has no beginning or end in time; it is finite but with no edges in space. He acknowledges that questions about the nature of the universe abound. Still, as he says, "Only time (whatever that may be) will tell."

that it ever will. Indeed, many of the trails the philosophers have followed lead only to a dead end.

SO, IF TIME DOESN'T FLOW like a stream, what is it "doing"? The best answer is that it's probably not "doing" anything, as B. F. Skinner said. "Physicists prefer to envisage time as all there at once," said Australian physicist Paul Davies, "a timescape stretched out in its entirety, like a landscape. It is a concept often referred to as 'block time.'"

For centuries, as we've noted, time was regarded as extending infinitely into the past and future and independent of the events that defined it. This implied that the simultaneity of events was absolute—separate from the situation of the observer. To put it another way, both Aristotle and Newton, who believed that time was absolute, thought that the duration of time between two events could be precisely measured, and that the interval would come out the same, no matter who measured it. Newton's position was especially forceful. In his monumental work, *Philosophiae Naturalis Principia Mathematica,* he minced no words: "Absolute, true, and mathematical time, of itself and from its own nature, flows equably without relation to anything external." Though brilliantly conceived by a brilliant man, Newton's concepts of time would be shown to be more metaphysical than scientific and outright wrong.

Despite the lapse, Newton's contributions to our current understanding of time were remarkable, notably his views on motion. Motion, as we know, is measured by time and space. Indeed, much of what we call motion smacks of time, and the very definition of motion indicates the relationship: the condition of a body when at each successive instant it occupies a different position in space. Velocity, a characteristic of motion, describes a change of position in time; acceleration, another characteristic, is

the rate at which speed increases with time. Motion in space, on the other hand, measures time—a 24-hour day based on one complete rotation of the Earth on its axis, a month by the revolution of the moon about the Earth, a year by one complete revolution of the Earth around the sun. We can also measure space with motion and time, using light-years, the distance light travels in a year, to determine the distance to faraway stars.

But equally important, Newton established that motion at a constant velocity was not absolute but relative. In fact, if two objects are speeding along relative to each other in a straight line at constant velocity, it is impossible to determine which is moving and which is not—unless there is some reference point to establish the difference. A familiar example of this is that of a passenger on a stopped train in a station. He looks out the window and sees a train alongside moving. As he fixes his gaze on the other train, he is unable to tell whether the train he's watching or his own is moving. But, by focusing on a reference object, the motionless station, the passenger realizes that the other train is moving and not his.

Since motion is relative and not absolute, it follows that it's impossible to measure any absolute, or real, speed—only relative speed can be measured. Everything in the universe is moving—the Earth, other planets, stars—so speeds can only be measured relative to something else that is moving. A person who feels at rest with ground beneath his feet might measure the speed of an apple falling "down" from a tree toward the apparently solid, unmoving ground. But what he has measured is relative. If the person could stand in outer space and observe all earthly phenomena, he would see movement in both apple and Earth, not just that of the apple.

Using another example of frame of reference,

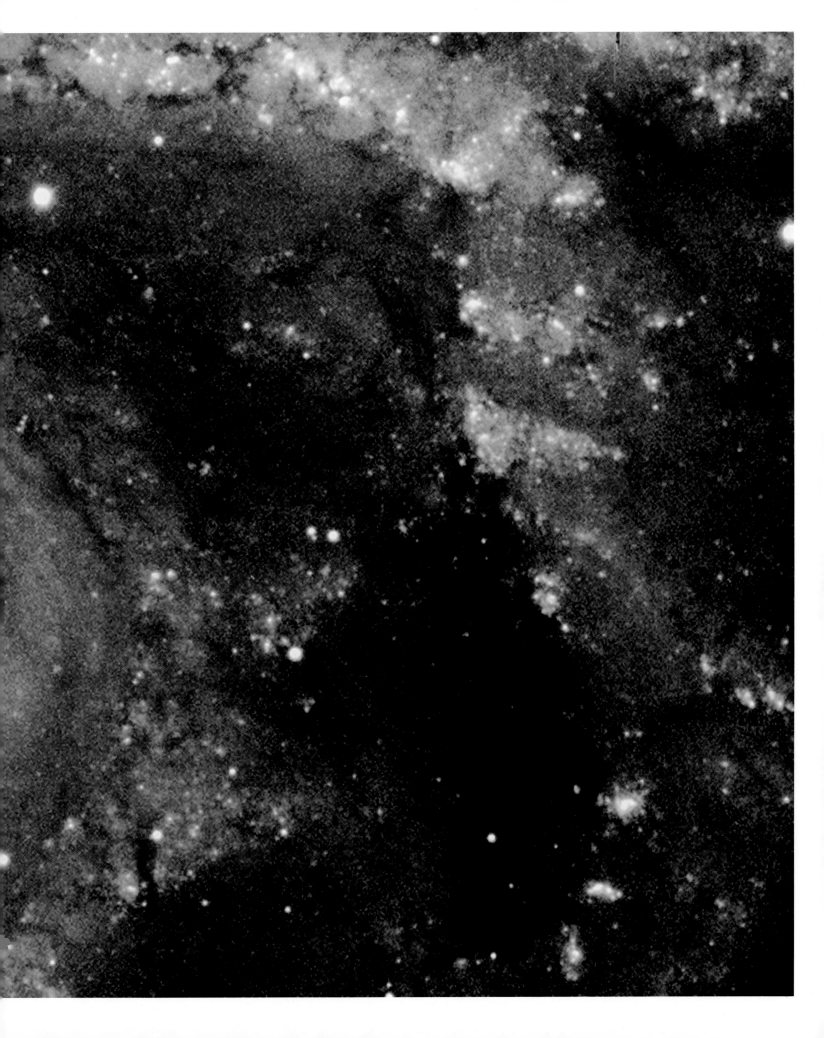

we've all experienced looking out that train window when it is moving, just as another train passes in the same direction. Even though we know we are moving forward, we get the feeling that we are moving backward, a phenomenon created by our momentarily shifting our frame of reference from the ground to the other train. And, strange as it may seem, both of the velocities we sense are correct—within their proper contexts.

Newton believed in absolute time, the flowing of which was "not liable to change…. All things are placed in time as to order of succession; and in space as to order of situation." It was the time of everyday, terrestrial experience. It took Einstein, with his own solid theorizing about the speed of light—essential to his theory of relativity—to wreck such notions and provide answers that have had far-reaching consequences affecting the concepts of time and space.

Where Newton had insisted that time, the absolute measurer of motion, was an independent entity that moved inexorably forward at a uniform rate, Einstein said no. Time was not absolute but was as relative as motion, a fourth dimension that becomes part of an a space-time continuum by which not only the where of an event could be determined but the when. What was absolute, according to Einstein, was the speed of light, something that in Newton's thinking was relative, since it depended on the observers' motion. The difference between the two views on light can be easily stated. Newtonian reasoning saw time was an absolute frame of reference, a "real" medium. Observers tracking a pulse of light beamed from point to point through this medium would not necessarily agree on its speed. But Einstein realized that the speed of light in empty space remains the same, no matter the movement and direction of the observer or the source of the light. Thus, it is absolute. It is always 186,282 miles per second, a universal speed limit,

blazingly swift enough to illuminate a cavernous stadium or auditorium or an entire city at the flick of a switch. The light from a star 50 light-years away will take precisely 50 years to reach an observer on Earth, whether or not the star is speeding toward us or away.

But while scientists agree on how fast light travels, they do not agree, in the reasoning of relativity, on the distance it has traveled, nor on the time it has taken. Stephen Hawking has explained what follows: "The time taken is the distance the light has traveled—which the observers do not agree on—

SIR ISAAC NEWTON:
Whether a falling apple
actually inspired the idea
of universal gravitation
in Newton's mind (the
story was first told by
Voltaire who got it from
Newton's stepniece) is
irrelevant. Newton's
ingenious contributions
to science laid the
foundations for much of
modern science; his
ground-breaking
*Philosophiae Naturalis
Principia Mathematica*
ranks as one of the
most influential books
ever written in physics.
The first scientist to be
knighted, Newton had
broad interests beyond
science, including
theology, alchemy, and
the chronology of
ancient civilizations.

divided by the speed of light, which they do agree on. In other words, the theory of relativity put an end to the idea of absolute time! It appeared that each observer must have his own measure of time, as recorded by a clock carried with him, and that identical clocks carried by different observers would not necessarily agree."

One way to demonstrate that time has no "right" clock is with one of Einstein's famed "thought experiments," mental gyrations that helped him crunch away at an especially nettling problem. Consider a man standing by a railroad track. He sees two bolts of lightning—one from the east, the other from the west—strike the same point in the track simultaneously. Understandably, he assumes the flashes occurred at the same time. Just as he sees the flashes, a train passes from west to east. How would a passenger on the train perceive them? Chances are he would think that the flash he is approaching occurred sooner because he is traveling toward it, and the other one later because the train is moving away from it. So, for one observer, the flashes are simultaneous, for the other sequential. Who's correct? Both, depending on their frame of reference.

Time, thus, is relative, not absolute. It cannot be regarded as measurable by an absolute succession of events independent of the observer. If one of the flashes was slower in coming than the other, it was not because time was actually moving slower in a real sense—it was only so relative to a frame of reference. Time has no real rate, no pace, no flow. Your now may be someone else's then. As Princeton physicist John Wheeler has observed, "Time cannot be an ultimate category in the description of nature. 'Before' and 'after' don't rule everywhere."

Sometimes, the term time "dilation" is used to explain an even odder phenomenon—a slowing down or stretching out of time. But for such an effect to be noticeable, motion has to approach the speed of light. Based on Einstein's calculations, this means that inside, say, a spaceship traveling at close to the speed of light, time drags compared to its passage as perceived by an outside, stationary observer; a clock aboard a spaceship traveling at 87 percent the speed of light would tick only half as fast as a clock on Earth.

In 1977, the theory was actually proved experimentally, though not at light speed, since no craft is yet able to fly that fast. Atomic clocks, incredibly accurate timepieces capable of measuring time to a few billionths of a second, were placed aboard an American satellite and sent into orbit. When they were returned to Earth and were compared to clocks at the Naval Research Laboratory in Washington, D.C., it was discovered that the moving clocks had, indeed, slowed somewhat, proving that Einstein was right: Time is relative.

Extrapolating from this, if a person were on the spaceship, would he or she experience a slowing of time, and, therefore, age more slowly? Theoretically and relatively speaking, it's not as far-fetched as it seems. Scientists know that tiny, short-lived subatomic particles called muons that have been created in the laboratory fall apart after a life span of only two-millionths of a second; but when they're zooming in from outer space at close to the speed of light, they survive much longer. Now consider the often cited example of twins. One of the twins born at the same instant travels in a spaceship at near the speed of light and is gone 20 years; the other remains behind on Earth. Aboard the spaceship, time passes half as fast as on Earth, but the space traveler doesn't feel the difference; to him, time is passing normally even though it has slowed up, relative to the ground, as his ship accelerates. The twin aboard the spaceship behaves and ages normally relative to himself and as measured by his own clocks,

ALBERT EINSTEIN, the man who made relativity and space-time household words, showed little promise of his later genius when he was a public school student in Munich. His slow speech development raised fears that he was retarded. But Einstein rebounded with a vengeance as a young man, publishing his special theory of relativity when he was 26 and his general theory when he was 37. He was 42 when he won the Nobel Prize in physics. A humanist as well as a physicist, he had a passionate interest in social justice and social responsibility, and, as a scientist, he disliked theories that left the fate of the universe to chance. "I cannot believe," he was fond of saying, "that God is playing a game of dice with the world."

which are in fact running slower than those on the ground. But his heart rate and respirations, say, would be normal, keeping pace with the clocks onboard his spaceship.

The theory goes that when he returns home, our space traveler will find that his twin has aged 20 years—the time the spacecraft was away—but that he, the space traveler, has only aged 10. He is, by earthly clock standards, younger; his twin is older, because he has aged at the relatively faster pace of Earth time. But such a scenario doesn't mean the space traveler's life span will be extended. He'd only live longer relative to his now older twin—not to himself. Thus, each twin has aged according to his own time.

But they might encounter a further complication. While everything aboard the spaceship has appeared normal to its occupant, as he rockets by the Earth at phenomenal speed, he'd get the impression that Earth clocks—if he could see them—were incredibly slow relative to his clock. Does this mean the twin aboard the craft is actually older not younger than his earthly twin? If that's true, when they reunite, each twin will be younger than the other, an obvious paradox.

In that intergalactic wonderland of time and light speed, things become curiouser and curiouser. Einstein knew that the mass (matter—or weight, in more common terminology) of a moving object increases as it approaches the speed of light. This has been adequately demonstrated in fast-moving subatomic particles—such as those that flash about in particle accelerators—and the results can be startling. The physicist Stephen Hawking, who holds Isaac Newton's chair as Professor of Mathematics at Cambridge University, tells us that objects moving at 10 percent the speed of light have a mass only 0.5 percent more than normal; at 90 percent the speed the object would be more than twice its nor-

mal mass. In today's particle accelerators, the mass of fast-moving subatomic particles increases many thousands of times, a phenomenon that requires enormous magnetic fields to contain the particles and keep them moving around a racetrack-like path. Even in an x-ray machine, the phenomenon is measurable: Moving electrons in the x-ray tube speed up enough to double their normal mass.

Subatomic particles and x-rays are one thing, but a spaceship is another entirely. As velocity moves toward light speed, the ship's mass increases, and as it rockets along faster and faster, even more energy is required to keep it accelerating. That energy, in turn, increases mass, which requires even more energy and so on and on until finally, at light speed, the ship would have an infinite amount of mass; that is, it would be limitless. An infinite amount of energy would then be required to sustain the acceleration of an infinite mass, something not in any immediate realm of possibility—nor even easily imagined by us earthbound humans.

Given the limits of human reasoning, and in light of all we've discussed so far, it might be advisable to face one simple fact about time: It may well be too abstract, or perhaps too basic, to be truly understood and explained. As the British scientist J.B.S. Haldane said of the universe, time "may not only be queerer than we imagine, but queerer than we can imagine."

It probably won't harm us one bit if we just continue to believe that time is there (or here) if only as a concept, or as a point or period when something occurs—perhaps only as a persistent illusion, as Einstein characterized the past, present, and future. What we know for certain, and what we will see in the next chapter, is that time is something we can use to tag events, to record and measure the rhythms of the spheres that govern our days and nights, and those within everything that lives.

CHAPTER 1

CYCLES OF NATURE

> "Wherever anything lives, there is, open somewhere, a register in which time is being inscribed."
>
> — HENRI-LOUIS BERGSON

EONS BEFORE philosophers and scientists existed to probe the mystery of time, before clocks and calendars became time's handmaidens, time moved here on Earth, hidden in the rhythms of nature, in the changes that would eventually signal its presence. It showed up first in the motions of celestial spheres, including that of this planet, in the changing seasons and ocean tides, and in the rocks that formed when the crust solidified, a record in stone of the changes the Earth has undergone through the eons. Later, as life appeared, time wound a new rhythm inside everything, from marine microorganisms to dinosaurs, from plants to humans—an internal clock as invisible as time itself, governing the way all living things function and die.

It is the same today. One spin of the Earth on its axis and an Earth day still goes by. One Earth circuit around the sun and a year passes. Our moon's apparently shifting shape as it travels around the Earth marks out a lunar month, and, with a little help from gravity and other heavenly bodies, raises the rhythmic ocean bulges known as tides. Deep within ourselves, time still ticks away the days, months, and years. And, even as we read this, nature continues to shift and reshape our Earth through such violent upheavals as earthquakes and volcanic action, proof that everything existing in time leaves traces of its changing condition, traces by which the past may be dated.

Time wears many faces as it runs through nature, not just, as we shall see in more detail, the expressionless countenance that stares back at us from clocks and watches or blinks electronically from a glowing digital display. And although all of time's aspects—its steady rhythms in heavenly spheres and in earthly biosystems, its regular patterns in geologic strata—are part of the same dimension, each has its special role in nature. The varying aspects of time touch in some way everything that exists, from a living organism to a layer of rock, both of which need the recurrent, measurable impact of time if they are to be put into their proper order.

Time, then, is embroidered not only in our human consciousness and in our cultures but in our very beings and the things that surround us. Its

SILHOUETTED BY THE AFTERGLOW of a setting sun, a darkened sea stack along the coast of Washington State looms imposingly over the tides that wash its base. The tiny crescent moon hovering placidly in the sky above serves as the timekeeper of those tides. As the moon orbits Earth, its distance from us shifts, and its gravitational pull varies, in turn causing variations in the times of the tides.

measured beat is everywhere. To understand how we and all living things function, and what the makeup of the Earth and indeed the rest of the universe is, we need time as a guide.

SOME 500 YEARS BEFORE CHRIST, the Greek philosopher and mathematician, Pythagoras, purportedly formulated a unique astronomical theory, based on the idea that the pitch of musical notes depends on the rapidity of vibrations. He believed that there was a mathematical relationship between the notes on a scale and the lengths of, say, vibrating strings sounding in mutual harmony. Extending this to the motion and distances of the planets, he suggested that, since the planets are separated by intervals corresponding to the relative lengths of strings producing harmonic tones, the planets must make sounds as they move. These sounds, which correspond to the planets' differing rates of motion, produce an ethereal musical harmony.

Pythagoras's ideas explaining all matter and its interrelationships numerically often were more mystical than scientific, but there is unquestionably a rhythm in the heavenly bodies that can be understood mathematically. While it may not result in increases and decreases in octaves that hang on planetary distances, it does impose a regularity, a repetition of events that can be seen and counted and, ultimately, used to measure invisible time.

Our time, which is to say Earth time, begins with two exercises in motion that, put in human terms, would be something like doing a continual pirouette while simultaneously and precisely circling, say, a fountain in the center of a park. Spinning on its own axis, the Earth swings on a year-long, oval-shaped trip around the sun, which itself is moving at approximately 67,000 miles an hour. As the Earth revolves counterclockwise around a tilted central axis, the North Pole then the South Pole slant

toward the sun, thereby varying the length of daylight during a year at various times and seasons on the Earth. But no matter how much daylight strikes the Earth while it is facing the sun—at the Equator there is no appreciable daylight change—one complete Earth revolution around the sun is what we know as a whole day, from morning to night and back to morning.

Since every day is a new one giving rise to a new tomorrow, it must have a beginning and an end. For us here on Earth, each calender day begins at midnight at an imaginary, zigzag line—the International Date Line that passes from the North to the South Pole through the Bering Sea and New Zealand at 180° longitude—and ends there as the Earth completes its rotation. Moreover, as the Earth rotates, the new date travels westward. So, when it's Sunday on the east side of the date line, it's Monday in the west. Thus, a traveler going from east to west across the line "loses" a day as he heads into tomorrow, and must set his calendar ahead a day; traveling east over the line he must set his calendar back, because he is moving into yesterday and "gaining" a day.

Relative, too, is the way an Earth day is calculated. There is solar time, which divides the day into 24 hours based on the regular passage of the sun across the same meridian: When the sun is highest in the sky at that meridian, it is noon, and the next pass of the sun over it marks the end of the day and the beginning of a new one. But Earth's skewed axis causes orbital fluctuations that bring it nearer to the sun. To complicate things further, the sun's own relative speed and position change. The day, then, varies in length, depending on the season, by as much as 16 seconds. For consistency, mean solar time was introduced, an artifice that relies on an evenly moving, hypothetical sun that allows for an average solar day of 24 hours.

Since mean solar time is based on the movement of a hypothetical sun, another form of measuring the day—sidereal time—is often used by astronomers. Instead of using the sun as a direct reference point, sidereal time uses an imaginary point in the sky, occurring at the vernal equinox, March 21. (Equinoxes are the two times a year when the sun's center crosses the Equator, and day and night are equal. The other, the autumnal equinox, falls on September 23.) A sidereal day has almost 24 sidereal hours, each with 60 sidereal minutes and seconds—but in total it's only 23 hours, 56 minutes, and 4.09 seconds long, which makes it shorter than the solar day. The discrepancy in hours carries over to the year, more noticeably since the Earth is revolving around the sun while it turns on its axis. In mean sidereal time, the Earth returns to the vernal equinox every 365 days, 6 hours, 9 minutes, and 9.54 seconds—in effect, the time it takes the Earth to return to the same place in its orbit with reference to fixed stars; in solar time, the journey takes 20 minutes and 24.04 seconds less.

Into this number-crunching timescape of moving heavenly bodies, yet another kind of time asserts itself. It is known as ephemeris time, a more precise, uniform measure based on the Earth's annual journey around the sun. Ephemeris time came into being because of the Earth's sometimes irregular and unpredictable rotation, which can deviate by a second or two a year. Moreover, the Earth experiences a gradual slowing down of about a thousandth of a second every hundred years, a phenomenon with no discernible impact on our daily lives but one that can be cumulatively significant. Ephemeris time takes into account both the moon's passage and variations in the movement of the Earth and moon, converting them mathematically into mean solar time.

Add to all this the fact that a Martian year is 687 Earth days long and that the planet Mercury rotates so slowly its day is longer than its year, and you soon realize that, while time can be measured with enormous accuracy, it is not a physical constant. We may be able to say that a day is a day is a day, and a year is a year is a year—more or less.

THE MOON'S RELATIONSHIP TO TIME measurement crops up in a more accessible way than in ephemeris calculating. Its position in the sky and its changing face from full orb to a glinting sliver as it passes around the Earth is a well-known phenomenon of nature, one that has long determined how humankind measures its months. Its gravitational effect on the Earth's ocean tides, a periodic rise and fall—roughly 12 hours between successive high tides—gives us one perspective. But because the moon's distance from the Earth fluctuates as it orbits, the gravitational attraction also varies in strength, and so, too, do the swells and times of the tides. Moreover, the timing of the moon's phases affects the height of the tides. When the moon is new or full, the tides are greater than average—so-called spring tides—because of the sun's extra, although lesser, influence; when the moon is at first and third quarter, the forces of the sun and moon oppose one another, resulting in lower, or neap, high tides. Spring and neap tides occur twice during a lunar month. Couple these fluctuations with the changing effects of the Earth's rotation and its own revolution around the sun, and tide time becomes a complicated thing. While the average break between high tides is 12 hours and 25 minutes, that interval can vary from less than 12 to more than 14 hours.

The predictable amount of time it takes for the moon to revolve on its axis and around the Earth also sets up a spin-orbit synchronization between the two bodies. This means that as the moon turns, we always see the familiar, apparent change in its shape only on one side, sometimes lit by the sun, sometimes dark. A full moon actually refers to the period when the whole apparent disc is illuminated, while a so-called new moon occurs when the dark part of the moon facing us just begins to be lit by the sun; both events are totally predictable in time.

Though it all sounds like perfect timing, there is a problem. We all know that the moon's passage around the Earth determines our month, but how long that is depends on how the trip is logged. For example, an anomalistic month—the time the moon takes to revolve from one perigee (its closest point to Earth) to the next—is 27 days, 13 hours, 18 minutes, and 33.1 seconds. But when the moon is viewed from a distant star, the complete revolution appears to take less time: 27 days, 7 hours, 43 minutes, and 11.47 seconds. Another way of seeing it is through the eyes of synodic time, the period between one new moon and the next, which averages 29 days, 12 hours, 44 minutes, and 2.78 seconds. Not exactly as indefinite as a month of Sundays, but astronomical proof of the importance of the eye of the beholder.

IT'S BEEN SAID THAT TIME is only what a clock's moving hands say it is. True to a point, but when it comes to geology, time is a stopwatch, fixed at a certain point to become the meticulous recorder of the sweeping changes that have long been a part of our natural history. Geologic time is a time machine on a grand scale, covering the millions and billions of years of the history of the Earth itself. Unlike the time associated with changes in position and movement of the heavenly bodies, geologic time needs no rotational regularity, no rhythmic waxing and waning to measure change. Its rhythms are played out and then frozen, embedded in the vast panorama of nature itself.

Change that occurs in nature over recent time is relatively easy to study, since often all that is required is dealing with the events of a paltry few years and centuries. These have typically been logged by a calendar set down by human hands, either in stone or in cuneiform characters, on papyrus or in printed books. Devastating earthquakes, volcanic eruptions, killer cyclones—each had eyewitnesses able to provide evidence of a natural event that became past, and the means to assign it a date, a time. But written records cover only a fraction of the Earth's history. To understand the greater, more ancient part of it and all the changes Earth has undergone over time needs digging, literally as well as figuratively. Detecting the changes, the record of past events stretching back millions of years, requires a time scale, a scientific standard, based on geology—not on clocks and calendars. Such a scale does not always conform with religious beliefs. For example, to some faiths, time and continuous change have no place in the nature of things: The surface of the Earth and everything in the skies around it were created "in the beginning," whole and unchangeable, by the hand of God. Fossils were variously considered to be "sports of nature," the results of nature's bungled attempt to transform the inorganic into the organic. "We must not forget that in the sixteenth century one was likely to lose one's life if one disputed the authority of the Bible," Bernhard Kummel, a geology professor at Harvard University explains of past beliefs. In that period fossils were assumed to be hardened lumps of matter "fermented by heat or by a 'lapidific juice.'"

Over the centuries, such traditional beliefs were challenged by mounting evidence that fossils were organic, signs of past life on Earth. As more and more preserved and recognizable remains surfaced—bones, shells, invertebrate animals, leaves, and dinosaur tracks—there could be no doubt that these were the remnants and impressions of life long gone. The surface of the Earth itself was not

SLICED BY WATER AND WEATHER, a Utah canyon bares a sequential record of the past. Such vertical panoramas reveal chronological layers of rock, formed by water-deposited sediments and plant and animal fossils. Prior to the late 18th-century, such geologic evidence was often cited to prove certain theological doctrines, but today scientists use radiometric dating techniques to analyze the layers and identify ancient life and changes in climate during specific geologic time periods. Using strata history, they can even estimate the Earth's absolute age.

static but swept by constant change. Even Aristotle knew this. "The distribution of land and sea in particular regions does not endure throughout all time," he observed, "but it becomes sea in those parts where it was land, and again it becomes land where it was sea. As time never fails, and the universe is eternal, neither the Tanais nor the Nile can have flowed forever."

But the evidence of Earth's lengthy history told early natural historians next to nothing about the age of the planet nor that of the ancient fossils buried in it. Scientists felt they needed a better way to date the Earth, along with its scattered rocks and fossils, at least on a relative scale that would tell

them which one was younger or older than the other. No conventional clock or calendar could help, and there was nobody around 4.6 billion years ago to take notes when the Earth was forming by planetesimal accretion, as the geologists say, or afterward, when its crust solidified to form the oldest rocks. Eventually, geologists and archaeologists found a way to tell "big time," or geologic time, the kind of time that takes over where minutes, hours, and the concept of a year fail miserably. More a calendar recorded in stone than a clock, it was a form of "writing" that could be read, provided one translated it into what would come to be known as geologic time. *Continued on page 42*

FOR MILLENNIA,
12,057-foot Semeru
has loomed over the
landscape in what is
now Java. And for cen-
turies, farmers have
terraced the volcano's
fertile slopes with fields,
ever wary that time will
one day run out, and
eruptions will bury their
crops under flowing
masses of molten earth.
With nearly 130
volcanoes dominating
Indonesia's skyline and
forming its backbone,
seismic unrest regularly
destroys fields—at the
same time replenishing
them with rich nutrients.

Earth–4.6 Billion Years Old

SUNDAY	MONDAY	TUESDAY	WEDNESDAY	THURSDAY
			45 DAYS AGO Earth forms by accretion; oldest known meteorite	**44 DAYS AGO** Accretion begins to slow
EACH DAY = 100 MILLION YEARS				
41 DAYS AGO	**40 DAYS AGO** Over 4 billion years before the present	**39 DAYS AGO** Era of great impacts ends	**38 DAYS AGO** The Iron Catastrophe destroys most of Earth's original crust	**37 DAYS AGO** Oldest known rocks on Earth form (found in Greenland)
Heavy meteor bombardment continues				
34 DAYS AGO	**33 DAYS AGO** Lava floods much of moon's near side, creating its "seas"	**32 DAYS AGO**	**31 DAYS AGO** Almost all of moon's present-day surface now formed	**30 DAYS AGO** Over 3 billion years before the present Small bits of continental
27 DAYS AGO	**26 DAYS AGO** Main continental cores forming	**25 DAYS AGO** Stromatolites increase	**24 DAYS AGO** Archean eon ends, Proterozoic begins	**23 DAYS AGO** First known glacial periods
		"Continental threshold:" first stable continent-size landmasses Crust now thick enough for high mountain ranges		
Photosynthesis by blue-green algae slowly adds oxygen to the environment				
20 DAYS AGO Over 2 billion years before the present	**19 DAYS AGO** Possible asteroid strike in Ontario leaves world's largest nickel source at Sudbury	**18 DAYS AGO** Stromatolites widespread	**17 DAYS AGO** Deepest layer of Grand Canyon forms	**16 DAYS AGO** First oxygen-dependent life
13 DAYS AGO	**12 DAYS AGO** Mountain building in eastern Canada	**11 DAYS AGO**	**10 DAYS AGO** Over 1 billion years before the present	**9 DAYS AGO** First sexual reproduction
		Early supercontinent breaks up into several different landmasses		
6 DAYS AGO First jellyfish; early glaciers retreat	**5 DAYS AGO** Paleozoic era begins; continents awash, very high sea level, explosion of sea life; first complex fossils; first seashells	**4 DAYS AGO** Sahara lies under an ice cap at South Pole; plants spread to land; North America and Europe collide; northern Appalachians rise	**3 DAYS AGO** Age of ferns, giant insects, coal forests; new ice cap in south; North America and Africa collide, southern Appalachians rise	**2 DAYS AGO** Pangea forms; Paleozoic era ends, Mesozoic begins; sea level drops; mass extinction in seas

Precambrian		Cambrian	Ordovician	•	Devonian	Carboniferous	Permian	Triassic

Silurian —

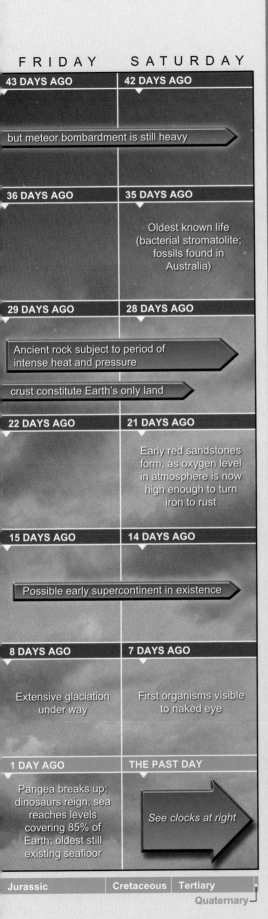

43 DAYS AGO | **42 DAYS AGO**

but meteor bombardment is still heavy

36 DAYS AGO | **35 DAYS AGO**

Oldest known life (bacterial stromatolite; fossils found in Australia)

29 DAYS AGO | **28 DAYS AGO**

Ancient rock subject to period of intense heat and pressure

crust constitute Earth's only land

22 DAYS AGO | **21 DAYS AGO**

Early red sandstones form, as oxygen level in atmosphere is now high enough to turn iron to rust

15 DAYS AGO | **14 DAYS AGO**

Possible early supercontinent in existence

8 DAYS AGO | **7 DAYS AGO**

Extensive glaciation under way | First organisms visible to naked eye

1 DAY AGO | **THE PAST DAY**

Pangea breaks up; dinosaurs reign; sea reaches levels covering 85% of Earth; oldest still existing seafloor | *See clocks at right*

Jurassic | Cretaceous | Tertiary
Quaternary

THE PAST DAY (present time is midnight)

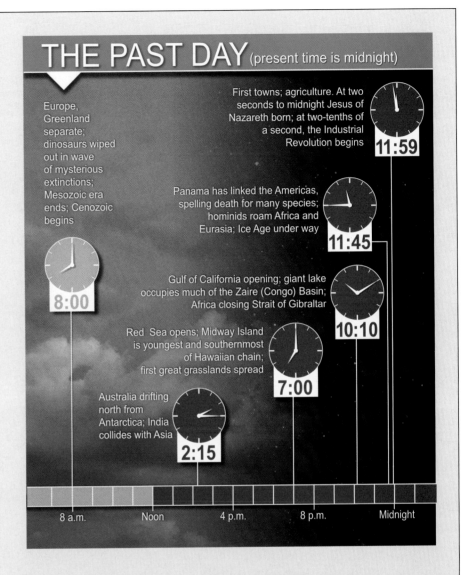

Europe, Greenland separate; dinosaurs wiped out in wave of mysterious extinctions; Mesozoic era ends; Cenozoic begins

First towns; agriculture. At two seconds to midnight Jesus of Nazareth born; at two-tenths of a second, the Industrial Revolution begins
11:59

Panama has linked the Americas, spelling death for many species; hominids roam Africa and Eurasia; Ice Age under way
11:45

8:00

Gulf of California opening; giant lake occupies much of the Zaire (Congo) Basin; Africa closing Strait of Gibraltar
10:10

Red Sea opens; Midway Island is youngest and southernmost of Hawaiian chain; first great grasslands spread
7:00

Australia drifting north from Antarctica; India collides with Asia
2:15

8 a.m.   Noon   4 p.m.   8 p.m.   Midnight

Imagine a calendar (left) in which each day represents 100 million years. On such a scale, the Earth, born about 4.6 billion years ago, would be 46 days old. A million years pass in less than 15 minutes, a thousand in just under a second. Like a film run on fast-forward, Earth is bombarded by meteors; its landmasses congeal into a massive continent, then gradually break apart; primitive life emerges and slowly evolves; forests and plant debris rot in primordial swamps to become coalfields; dinosaurs rule and are swept into the fossil record; ice ages come and go; and, finally, in the last moments on the last day, humans emerge and begin to excercise dominance on the planet. Theirs is a new era. Will they be as successful or last as long as the dinosaurs? Only time will tell.

READY TO BECOME A FOSSILIZED RELICT, an ant lies trapped in resin spilling from the trunk of a tree in a French Guiana rain forest. Fossilized resin, or amber, holds countless clues to the creatures that inhabited Earth's distant prehistory, preserving even the DNA of long-extinct flora and fauna.

SURROUNDED BY A FEATHERED AURA, the remains of 120-million-year-old birds are etched into a rock layer. Their size difference and the long tail feathers of the male (left) indicate that sexual dimorphism may have characterized bird life for eons.

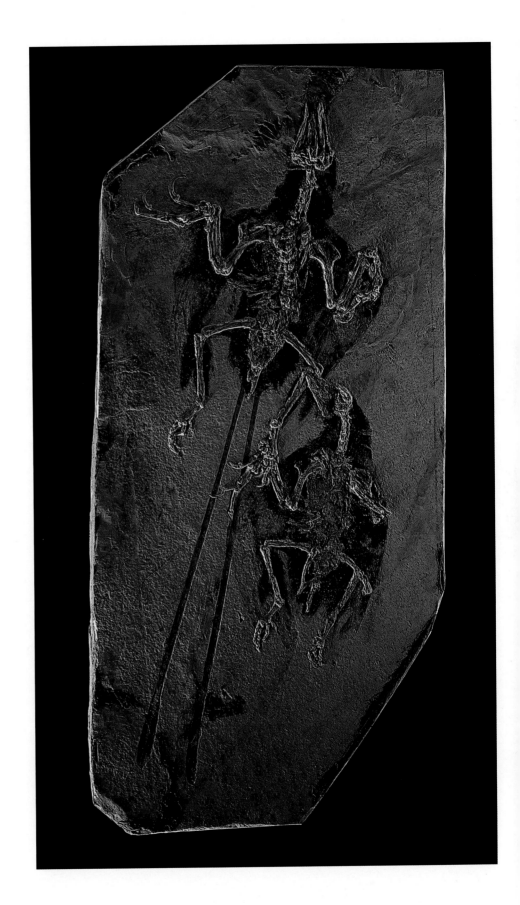

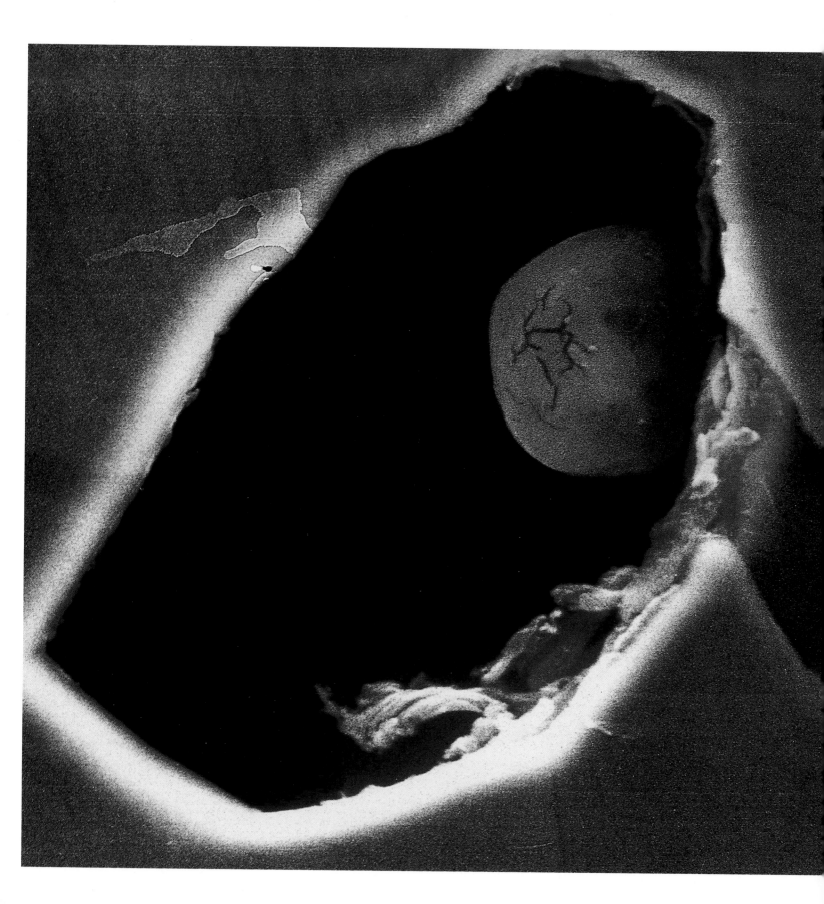

OLDEST EVIDENCE OF LIFE ON EARTH?

Scientists believe a tiny ball of carbon (opposite) nestled in the cavity of a 3.86-billion-year-old rock may be just that. Magnified here 6,900 times, the specimen was found on an island off Greenland in 1991. Though it has no anatomical features, scientists think its biochemistry was similar to that of every carbon containing life-form that has evolved since. At right, a scientist takes shavings from a fossilized reindeer bone, for radioactive carbon-14 dating. Since carbon 14 decays at a uniform rate, the extent of its decay in dead animal and plant remains allows scientists to date the remains and the geologic formations in which they were found.

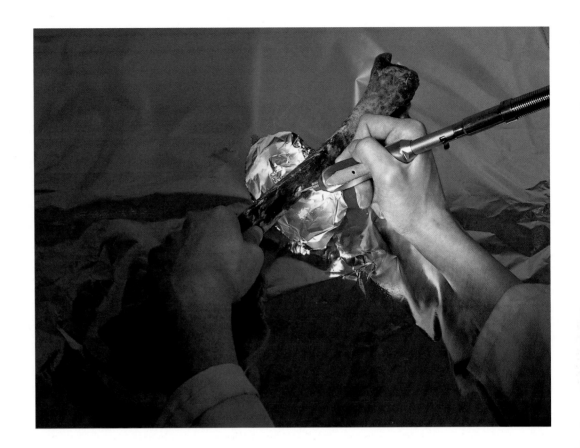

For those who know the language, the records of geologic time are in the exposed and weathered sedimentary rocks beneath our feet and in the massive horizontal rock layers, or strata, laid down from the time Earth was formed. It is in the mesas and buttes time has sculpted and scarred, in the precise striations of the Grand Canyon, in the continents set adrift and then anchored, in the baked beds of long-dry seas. Geologic time has left its traces, too, in the scars made when hurtling asteroids, unseen by human eyes, blew portions of the Earth's surface to smithereens; in the age rings in a cross section of an ancient tree trunk; and in the millennia-old desert dust blown from afar and trapped in a column of ice cored from an Antarctic glacier. Each of these is proof of nature's ever changing face and of time's role in the process.

Among the first to read nature's record book was a 17th-century scientist who followed a profession that seems, at first glance, to be far removed from geology. Niels Stensen (whose Latinized name was Nicolaus Steno) was a Dane, an anatomist who was also a theologian and a Catholic bishop. Whether it was from his observation that muscles were layered in bundles and parallel planes or his dissection of a shark's head (which indicated to him that the "stony concretions" found earlier in Tuscany were fossilized shark's teeth), Steno became intensely interested in the layering of sedimentary rock and in its relationship to fossil remains. His investigations provided scientists with an important fundamental concept of historical geology—the superposition of strata.

In studying the relative positions of sedimentary rocks, Steno found that the rock particles forming them—gravel, sand, and mud—settle, according to their size and weight, to the bottom of bodies of water. The largest settle first and the smaller, lighter ones last. The whole collection eventually forms rock layers piled one on top of the other after each has solidified, burying living and dead animals in the accumulation. It soon became obvious that in any sequence of layered rocks, a lower bed had to be older than any bed on top of it. Moreover, fossil hunters now had compelling evidence of organic evolution in the layers: older, simpler forms of life at the bottom changing gradually upward into more complex, younger forms. Steno's Law of Superposition, a simple but telling observation, had uncovered time's hiding place in the muck and rock of the Earth.

As valuable as Steno's stratigraphy was, it explored only the relative ages of rock layers and the fossils within them. An improvement was the geologic time scale, which developed over the 18th and 19th centuries. This divided time into convenient intervals, each containing the rocks deposited during the divisions. The longest intervals are eons, each divided into eras; eras are subdivided into periods, which are subdivided into epochs. But, like Steno's law, such geologic time scales are relative, and they say nothing about how long each division and subdivision actually lasted. To determine exactly how ancient all the rocks and fossils are, time had to be expressed in numbers of years, in what's known as evolutionary and geologic timelines.

Over the years, there were numerous attempts to estimate the Earth's absolute age. The Bible provided one source: In the 1650s, Irish archbishop James Ussher, using scriptural sources as well as astronomical cycles and historical accounts, determined that creation had occurred the evening of October 22, 4004 B.C. Other investigators relied on science. In the late 19th century, scientist John Joly used a different tack. Calculating the total amount of sodium salts in the world's oceans, as well as the amount added every year from rock erosion, he came up with 100 million years as the approximate age of the Earth. His figure was generally accepted for some

MOON TIME

For eons, humans have looked to the skies to chart the passage of time. The moon, Earth's only satellite, is a particularly telling timekeeper. About 240,000 miles away and perpetually circling the Earth, the moon takes roughly 27 days to complete one revolution. Because it rotates on its axis while revolving, it always presents the same face. Reflected light from the sun gives it its shine, and the different angles from which we see that bright, reflected surface cause what we call its "phases." The inner circle in the diagram below shows these phases as we see them from Earth; the outer circle shows the moon as seen from high above our North Pole. Because the Earth spins faster than the moon revolves, the satellite rises about 50 minutes later each night. During the new phase, moon and sun rise and set simultaneously. After that, the moon appears in different parts of the sky: in the west as it waxes larger toward gibbous, in the east as it wanes smaller.

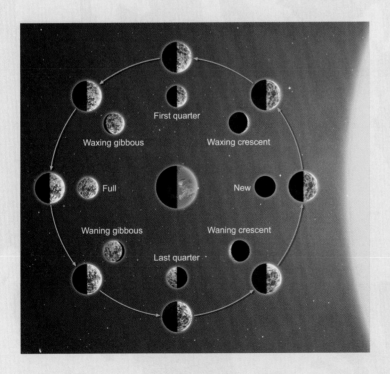

time, until later scientists realized that such a time frame wasn't long enough to account for the Earth's entire history and for organic evolution. Other methods measured the deposit rate of limestone and like materials, but such variables as erosion, rainfall, and sunspot cycles (which affect weather patterns) made the readings undependable.

In 1947 American chemist Willard F. Libby found a seemingly foolproof method of dating that went beyond simply identifying the age of ancient artifacts and fossils. By using the rate at which radioactive elements decay or lose their radioactivity, Libby's method could be adapted to fix the time when the Earth and the moon formed. Specifically, Libby focused on radioactive carbon-14 atoms, which are continuously produced in the atmosphere by neutrons bombarding nitrogen. With a relatively long half-life of 5,730 years (the time required for half its radioactivity to disappear), carbon 14 had the makings of being an accurate, ticking clock. It not only bathes the Earth, but its atoms react chemically with oxygen to form carbon dioxide. This, in turn, is absorbed by plants; animals eat the plants, also taking in carbon 14, and humans absorb it by eating both plants and animals. The end result is that all living things, as well as numerous objects, contain carbon 14 in the same concentration that occurs in the atmosphere during their lifetimes.

When a living thing dies, it stops absorbing radioactive carbon and what remains disintegrates at a measurable rate, reverting to nitrogen 14. By comparing the ratio of carbon-14 atoms to that of nitrogen-14 atoms, thereby calculating the amount of carbon 14 that remains stored in a sample, scientists can tell the age of a vast number of antiquities, from ivory and iron, to mummies, bristlecone pine trees, parchment, and bones.

The amount of carbon 14 left in a sample after

about 50,000 years is too small to use as an accurate measure. To date very old rocks and fossils, scientists turn to radioactive elements with longer half-lives: uranium 238, for example, with a half-life of 4.5 billion years; or potassium 40, 1.3 billion years. On a much smaller scale, relatively short-lived radioactive isotopes like tritium also double as clocks. A form of hydrogen, tritium has a half-life of only a dozen years and is used to study reaction rates in chemical analysis, to determine the rate at which rainwater moves through rock cracks, and to keep track of the age of wines.

But it is the long-lived elements that have finally given geologists an absolute time scale that affixes a number in m.y.a.—million years ago—to a given sample of rock and to each of the divisions of the current geologic time scale. For example, the Cenozoic era is now seen as covering 65 m.y.a. to the present, the Mesozoic from 245 to 65 m.y.a., the Paleozoic from 544 to 245 m.y.a. Indeed, as Emerson believed, "The years teach much which the days never know."

TIME'S ROLE AS A MECHANISM that controls life is evident everywhere. We see it most often in the natural rhythms that take us from day to night to day again; in the change of seasons and in the rise and fall of ocean tides. But we are governed as well by biological rhythms. Indeed, there are rhythms within virtually everything that lives, regularly repeated functions and behaviors that are often synchronized with the 24-hour, day-and-night cycle and with lunar effects and seasonal changes. It is as though the movements of the spheres have somehow been stamped into the genome. Unconsciously, we keep our own time.

But although biological rhythms may work in tandem with the day-night cycle, they are not necessarily directly controlled by it, and often they will continue to keep their own innate sense of time—for

a time—even when a sunrise or a sunset is absent. Such persistence, scientists believe, is proof of organic clocks, time-keeping mechanisms buried in the cells, each capable of driving or coordinating a rhythm and regulating processes within every life form.

Organisms of all shapes and sizes—plants and animals, mushrooms and bacteria—depend on such clocks, albeit clocks that are infinitely more complex than any hanging on a wall or wrapped around a wrist. Some of the rhythms they regulate work on a yearly cycle, others by the lunar month, and still others—in fact the vast majority—are set to a 24-hour, or circadian schedule. (*Circa* means "roughly," and *dian* refers to "daily," a point that will become relevant later.) One set of clocks and rhythms may control a plant's flowering, nectar secretion, periods of dormancy, and seed germination. Another is inside bees, birds, and butterflies, helping them navigate, migrate, propagate.

There are rhythms and clocks in every fiddler crab, signaling that low tide is imminent and that it's time to hit the beach; in every bear, letting it know when to hibernate; and in every mold, telling it when to release its spores. The clocks in raccoons and hamsters alert them to become active at night and to sleep during the day, while the clock in a sparrow works in the opposite way. Clams, snails, and other mollusks possess clocks that control the layers of materials making up their shells. These telltale rings are not unlike the annual growth rings in a tree trunk, which, through the technique of dendrochronology, we use to estimate a tree's age. Not the least are the biological rhythms and clocks in humans. Internal clocks time our metabolic systems, brain waves, blood pressure, heartbeat, digestion. They control our sleep-wake cycles, pulse hormones into our bloodstreams, and keep our reproductive rhythms rolling. In short, they keep us alive and pace the rate at which we age.

The way the rhythms themselves are generated varies. Some may be self-sustaining within a lifeform, arising through internal metabolic processes and set to control inherent periodic systems, such as brain waves and heartbeat. Such rhythms are determined simply by the time it takes to complete a specified sequence of actions—so many cycles per second for brain waves, for instance, or the periodic contraction and relaxation of the heart muscles. The rhythms may also be exogenous, that is, imposed on the bioclock by changes in the outside environment. Not only does the fiddler crab, for instance, emerge from its burrow at the surf's edge at low tide, but also its skin regularly darkens around sunrise and blanches near sunset—a pattern that seems to reflect both lunar and solar rhythms. Biologists can tell whether a particular rhythm is endogenous—generated within an organism—by separating it from outside influences and placing it in an environment with unvarying conditions.

Which is how the study of chronobiology—that is, the study of biological rhythms—began in 1729. Like Steno, who moved into geology from anatomy, Jean-Jacques Dortous de Mairan, a French astronomer, briefly shifted his attention from the stars to the Earth and in the process came up with the first demonstrable evidence of an internal timekeeping mechanism in a living organism.

SEARCHING FOR FROZEN CLUES

to the past, researchers examine the dust, gases, ash, and sediments trapped inside an ice core taken from a Swiss glacier. Such samples provide data on events that have impacted the environment. Cores taken from Antarctica and Greenland ice contain a chemical record of atomic-bomb testing, fallout from the Chernobyl nuclear accident, global climate changes, even the advent of atmospheric pollutants during the industrial revolution.

His object of interest was the heliotrope, a plant known not only for its fragrance but also for the rhythmic opening of its leaves at daylight and closing of them when darkness falls. Curious about what would happen if the plant were removed from the cycle of day and night and kept in total darkness, de Mairan put some heliotropes in a darkened shed. Even in the dark, the plants opened during the day and folded their leaves in the evening, keeping them shut until the appointed time for opening in the morning. "The sensitive plant senses the sun without being exposed to it in any way, and is reminiscent of that delicate perception by which invalids in their beds can tell the difference between day and night," a scientist reading from de Mairan's experiment reported to the Royal Academy of Paris.

Though the heliotrope was not actually sensitive enough to be aware of the sun without seeing it, it did have an internal mechanism—something not grasped by de Mairan—that matched the cycles of nature associated with the passage of time, even when the plant was isolated from its environment. Honeybees display similar signs of such a clock. In 1929 Karl von Frisch, an Austrian zoologist, saw that when bees were trained to visit a nectar feeding station between, say, 4 and 6 p.m., they would continue to visit between those times even when the nectar was removed and an outside clue, like light, was absent. In 1960, Karl C. Hamner of the University of California at Los Angeles imposed even more stringent conditions on his test subjects. He took fruit flies, hamsters, soybean plants, cockroaches, and bread molds to the South Pole, where he placed them all on a turntable set to rotate at exactly the speed of the Earth—but in the opposite direction. By doing this, he exposed his test subjects to an environment without any earthly indication of daily time. Despite the disruption, their regular activities continued to follow a 24-hour cycle.

Hamner's menagerie, along with bees and heliotropes, could tell time.

But whose time? When an animal's circadian clock is set to local time, it is in sync with, say, its prey, its sources of nectar, the sun by which it orients itself, and members of its social group. Removed from environmental time cues and kept in constant environments, animals keep their daily rhythms for a time, but eventually, according to some scientists, they display free-running time, a natural, self-sustaining rhythm independent of local time. (In humans, it's a period a bit longer than 24 hours.) Animals and humans on free-running time soon lose touch with the real world, and, while they may eat and sleep on a regular schedule, they are increasingly out of phase.

Such drifting doesn't occur under ordinary circumstances because external cues reset the circadian clocks every day, much as we reset our home clocks by moving the hands to another hour. With inner works driven by molecular and cellular structures, not pendulums and mainsprings, circadian clocks can often be reset simply by shifting the light-dark cycle familiar to a certain life-form. Resetting a bioclock usually requires a synchronizer such as light or temperature. The synchronizing effect of light, which biologists call an entraining agent, has been well documented. Plants can be made to flower or bear fruit by subjecting them to artificial light at night and keeping them in the dark during the day. Caged sparrows can be entrained to an altered, artificial light-dark cycle; birds that navigate by the stars can be made to orient themselves under the artificial sky of a planetarium; human beings can create new sleep-wake cycles by timed exposure to bright light or by going to bed a certain number of hours later every day.

Resetting the human clock is, however, not always easy. When we *Continued on page 56*

RITE OF SPRING:
In a Maryland pond, spadefoot toads mate in a synchronized seasonal ritual inspired by their inner biological clocks. Such temporary spring ponds, created by snowmelt or rainwater, act as spawning grounds for thousands of species worldwide.

"THERE IS A TIME FOR EVERYTHING and a season for every activity under heaven," the Bible proclaims. A resplendent row of autumn-burnished sumac leaves (above) confirms that, reminding us of the ephemeral nature of time's passage. The leaves' annual display, called "the color" by New Englanders, lasts only a few days, soon to be replaced by the advent of winter. Full of its own marvels, winter can etch a windowpane with patterns of sunlit frost (opposite) that

create an evanescent
beauty. Tropical palms,
too, must bend to the
will of seasons and time,
as they do (opposite,
lower) during a May
storm in Miami Beach.
Before clocks and calen-
dars guided humankind,
the repetitive pattern
of seasonal changes
gave evidence of time's
presence, its passing,
and its proclivity to
begin anew.

BIOLOGICAL CLOCKS,
ticking away in the cells
of creatures large and
small, mete out life
spans according to a
hereditary blueprint.
Tortoises, like this giant
on Galápagos Island
(opposite), routinely live
to be a hundred years
old, while mayflies
(right) may live for only
a day. Inner clocks do
infinitely more than
govern how long we
live, however. Set to a
roughly 24-hour, or
circadian, schedule, they
regulate functions from
sleeping and waking to
heart, brain, and meta-
bolic functions. Present
in plants as well,
internal clocks control
flowering, dormancy,
and such daily activities
as leaf movements and
nectar secretion.

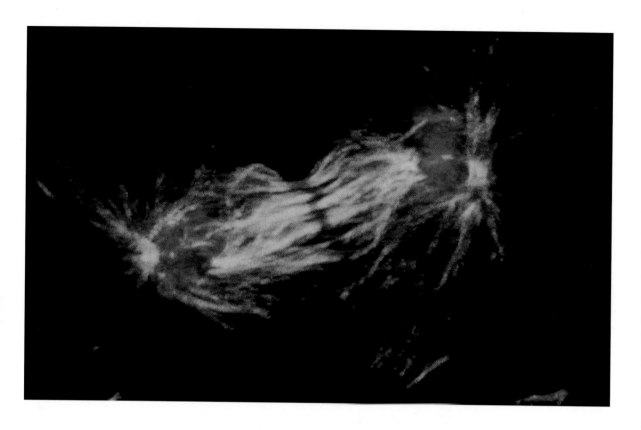

an animal cell (left) divides in the precise biological process known as mitosis. The cell's nucleus forms two new nuclei, each with the same number of chromosomes as the original. The number of cell divisions each organism undergoes is proportionate to its life span. Research on fruit flies indicates the DNA molecule responsible for programming their biological potential is a "clock gene." Opposite, a laboratory worker charts human DNA, our link to past and future. As ongoing DNA studies map our genetic universe, they show promise of leading to cures for disease and answers to the mysteries of evolution.

cross numerous time zones—including the International Date Line—during, say, a 14-hour flight from Tokyo to New York, we are way out of phase, because of the time difference between our body clocks and local time. Our watches tell us what time it is in New York when we arrive, but our circadian clock is still on Tokyo time. Disoriented physically and mentally by the warping of time known as jet lag, we desperately need sleep, even though local time and sunshine tell us we should be starting the day. Often, several days are required before our clocks are reset, allowing us to readjust to a new sleep-wake cycle. Our inner clocks are not unlike those that allow insects and birds to find their way across great distances and to adapt simply by the presence and angle of the sun.

But what exactly is the biological clock and where is it? If it is a physical entity, one that we inherit in order to keep time with our surround-

ings, what genes control it? Does the time the inner clock keeps coincide exactly with that measured by the rotating and revolving of the Earth?

The answers are as difficult to come by as explanations of time itself. However, scientists are beginning to get a better grasp of biological clocks as they examine our organic gears and springs and the rhythms that drive them. The consensus is that organic clocks are inborn, not learned, and need to be triggered to keep proper time. Also, the location of the clock varies with each life-form and is present even in different parts of individual cells. Constructed of tissues, proteins, and genes, the organic clocks that power circadian rhythms are complex mechanisms whose chemical reactions control the timing of a wide range of events, including cell division. They have been found in the brains of humans; in the optic lobes of cockroaches, crickets, and silkworms; in the eyes of mollusks;

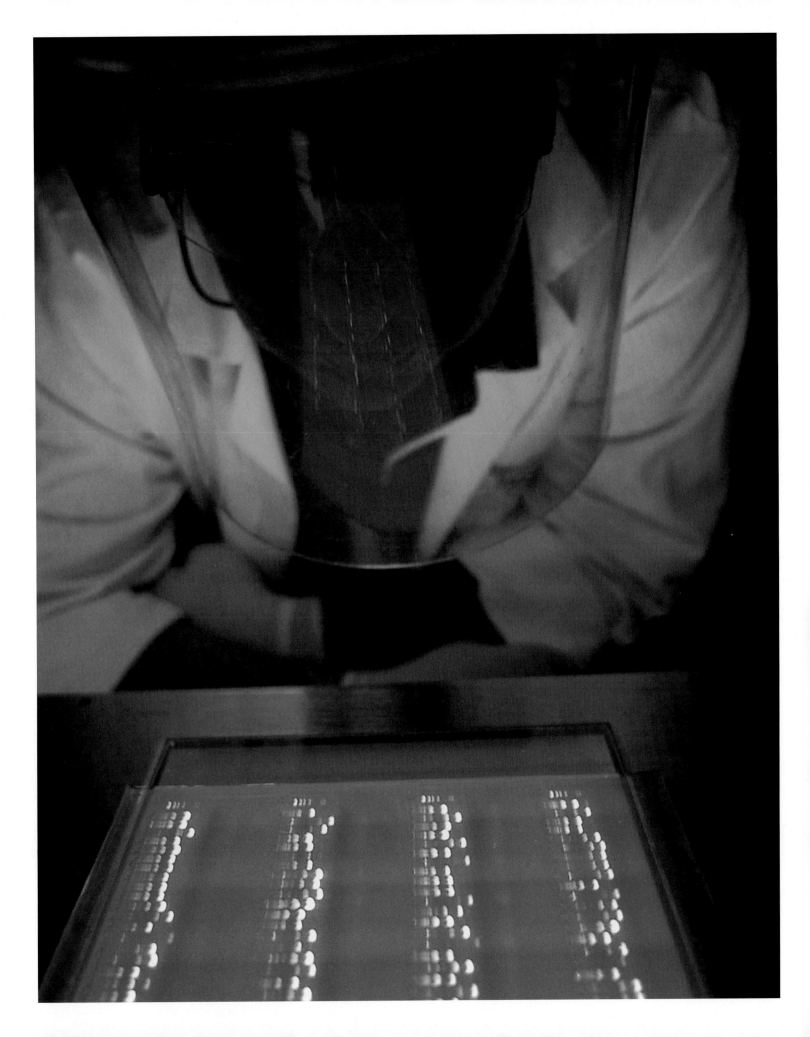

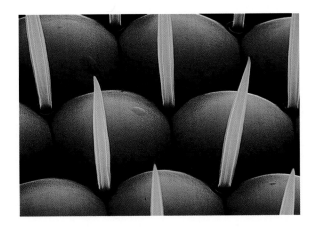

and in the thoraxes and abdomens of fruit flies. There are even clocks in the genetic structure of cyanobacteria, the simplest organisms known to have internal clocks: When their clocks are working correctly, the organisms can be made to glow in pulsing rhythms, like fireflies. These clocks are in all living cells, causing them to double in proportion to life spans—12 times for a mouse, 25 times for a chicken, 50 times in a human living 115 years.

That clocks exist and run independently in tissues outside the brain was demonstrated not long ago by cell biologists working with "clock genes" in fruit flies. Earlier work had identified such genes, appropriately named "per" for "period" and "tim" for "timeless." At the time, it was generally believed that, while all the flies' cells contain the per and tim genes, it was the brain cells that set the decisive pace for the flies' clocks. Recent research, however, suggests that that may not always be the case. In an effort to determine whether individual body parts of fruit flies would respond to changes in the light-dark cycle, cell biologists at Scripps Research Institute in San Diego fused the flies' clock DNA to genes that give the "glow" to jellyfish and fireflies, thus creating glow-in-the-dark fruit flies. With changes in external lighting, the flies glowed on and off rhythmically, notably in the chemosensory cells

at the base of hairs on their legs, wings, antennae, and proboscises. "Our findings confirm that body clocks run independently in many tissues outside the brain," said researcher Steve Kay, "and are reset by light, implying that cells harbor novel photoreceptors that aren't involved in vision."

Although this research challenges theories of the brain's role as the coordinator of circadian rhythms, it's not to be discounted by any means, not even in fruit flies. In the Scripps experiment, the fly brain was the only organ in which the clock genes stayed in sync in the prolonged absence of light. Still, the many nonbrain clocks in fruit flies raise the intriguing possibility that the skin, liver, and other tissues of humans may have their own clocks capable of controlling local function. Thus far, no tim or per genes have turned up in humans (clock genes have been identified and cloned in mice), and the consensus among scientists is that in humans, the clock is in the brain, specifically in the structure known as the suprachiasmatic nucleus (SCN).

This human circadian pacemaker is a tiny cluster of some 10,000 brain cells situated just above the optic nerve in the hypothalamus, the master controller of the endocrine system, the autonomic nervous system, and such bodily states as hunger, thirst, temperature, and emotions. First identified as a circadian clock in rats, the SCN is but one gear in the mechanism that keeps biological time. Light receptors in the retina of the eye lead to the SCN, which takes information on day length, interprets it, and passes it to the pineal gland. This pea-shaped gland is situated behind the hypothalamus, and, although all of its functions are not fully understood, it secretes the hormone melatonin in reply to stimulus from the SCN. Melatonin interests biological-time researchers, because nighttime causes the hormone secretion to rise, while daylight inhibits it. When light is absent, melatonin is still released rhythmically, but when the SCN is destroyed, circadian rhythms disappear entirely.

Even though our inner clocks seem able to tell the time of day, and even though 24 hours is nature's spin and underlies many of our functions, does this mean that our inner clocks always reflect what is going on outside? Exactly how long is the circadian day anyway?

SCIENTISTS HAVE BEEN speculating on the precise length of human daily rhythms for years. It may be fine for us to say "it's about ten o'clock" in passing conversation but that kind of inexactness doesn't suit scientists—especially those studying sleep disorders and biological clocks. "About," however, might not be so bad a word, considering that a sidereal day is shorter than a solar day, that free-running time is longer than 24 hours, and that besides circadian rhythms there exist ultradian rhythms. These biological rhythms, present within the sleep cycle, have a frequency of less than 24 hours. Plants also seem to have inconsistencies in their cycles: Seedlings in the same garden may exhibit different behavioral patterns, raising and lowering their leaves, say, in 23- or 25-hour cycles.

There's no question our sleep-wake cycle is inconsistent, idiosyncratic, and flexible. Babies wake up and go to sleep when they want to, not when their parents want them to. As near to random as it can get, the sleep cycle of infants doesn't really pick up the circadian tempo until around six weeks of age; it takes some sixteen weeks for full entrainment, a process that may be influenced by an important social cue, the parents' schedules. On the adult side, our often hectic work and social schedules, jet travel that fatigues and disorients us—not to mention quantities of coffee and other caffeine-containing drinks—all conspire to gum up our biological clockworks and upset the normal sleep-wake cycle.

Still, most of us continue to tick along, adapting, picking up on the new rhythms as easily as we would play a tune by ear. We shift to and from daylight saving time without missing too much of a beat (in musical terms again, it's sort of like not noticing when a concert pianist hits one wrong note), and it's doubtful that our night-day, sleep-wake rhythms suffer because there's an extra day in a leap year. We can even cope with drastic changes, like 21- and 28-hour days imposed experimentally to study time adaptation.

So why worry about the exact length of a circadian day? Other than science's inherent need to know, there are several practical reasons. If, for instance, a worker's job involves shift rotation, his or her biological clock can't be reset by, say, exposure to bright light or changing sleep patterns until we have established an accurate measurement of a circadian period. Jet lag, the difficulty astronauts have sleeping in an orbiting space station, and even winter blahs and spring fever can all be dealt with more effectively if internal rhythms can be pinpointed. So, too, can patients suffering from illnesses whose symptoms are in sync with a circadian rhythm. It's fairly well known, for example, that medications for asthma, epilepsy, cancer, cardiovascular disease, and allergies work better and with minimal side effects when given at particular times of the day.

For years, the consensus among scientists was that the internal clock seldom ran for exactly 24 hours. It was more apt to drift toward 25 hours, unless it was set back an hour each day by exposure to morning light and to external clocks. This was demonstrated in the 1960s, in studies where test subjects stayed in an isolated bunker for several weeks. Kept from external cues but allowed to turn lights on and off at will, the subjects developed a sleep-wake cycle of about 25 hours.

But recent research suggests that the human

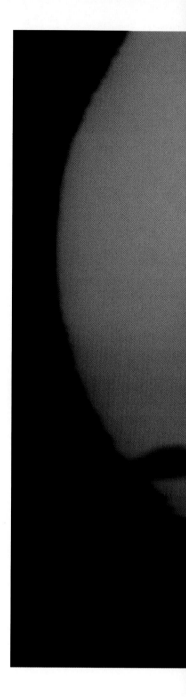

GUIDED BY INTERNAL biological chronometers, a flock of snow geese (above) migrates over Iowa. The annual movement of birds back to a breeding area may be influenced by an inner timing mechanism that alerts the birds to seasonal

weather changes in advance. Migrating birds apparently use many environmental clues to navigate—the sun and stars, geographic landmarks, coastlines, even wind direction.

cycle may be closer to what the timekeeper of Earth's rotation has always said it was: 24 hours. In perhaps the most accurate measurements thus far of human-activity rhythms, Harvard University researchers concluded in 1999 that our internal clocks run on a daily cycle of 24 hours, 11 minutes. To arrive at this conclusion, they "disconnected" test subjects from clock time by having them go to bed four hours later each day—essentially creating a 28-hour day. This strategy meant that the subjects (24 men and women, young and old) were on a

cycle that distributed light exposure, sleep and wakefulness, work and play evenly around the biological clock. Instead of now getting light exposure at the same time each clock day, the subjects experienced a six-day week in which light and dark occurred at different times each day. This revamped schedule freed internal clocks from the sleep-wake cycle, enabling them to tick at their natural pace.

Despite six-day weeks, body temperatures and hormone secretions went through seven cycles every week; no matter when the subjects retired or woke, no

WILDEBEESTS KICK UP CLOUDS of dust in Zambia's Liuwa Plain National Park (top), while migrating reindeer (bottom) move through a winter white world in Norway. Many animals have the instinctive ability to migrate hundreds, even thousands of miles, sometimes through unfamiliar territory. While the processes that lie behind such innate navigational prowess are not fully understood by humans, the regular movements of birds, mammals, and fishes have long given us a way to record the passage of time and prepare for changes in the seasons.

matter what they did while awake, their body temperatures and hormones rose and fell on an average cycle of 24 hours and 11 minutes. There was another significant conclusion: The human circadian pacemaker was now seen to be as stable and precise in measuring time as that of other mammals, many of whom appear to have a more finely tuned biological clock and a more instinctive time sense than humans.

Accepting the near-24-hour period suggests that all the old notions about daily human rhythms we have taken for granted have to be rethought. "Biological clock lore states that we drift to a later wake-up hour on weekends because we fail to reset the 25-hour cycle each morning as we go to work," said Charles Czeisler, the professor of medicine at Harvard who led the study. "We're not drifting. We're pushing ourselves to a later time with our exposure to electric lights from sunset to bedtime. That resets our biological clocks." If this resetting leads to difficulty in getting up on Monday morning, what can we do about it? Czeisler had an answer for that: Go to bed earlier on weekends.

The complexity of a biological clock that can be set and reset may one day be explained more fully. What can be said now with some certainty is that the Earth's rotation has somehow impressed a 24-hour period on the genes of living creatures, and that, although the biological clock is innate, it is also responsive to, and strongly influenced by, various environmental conditions. If our origins are indeed in stardust, then we are all necessarily connected to those cycles of nature that arose in the cosmos. And connected, too, to the time that is within and essential to all of life.

LIGHT ON THE SUBJECT: Scientists have long understood that light greatly affects the circadian clock, and new research indicates that shining a light on the back of the knees (below left) may alleviate jet lag or insomnia by sending a message to the brain via blood flow. At Cornell University Medical Center, scientists used volunteers to chart their production of melatonin —a regulator hormone usually released in the evening—along with their lowest body temperatures, usually occurring before dawn. After light was applied to the knees, the volunteers' body clocks shifted as much as three hours. At right, a beam of light shines in the eyes of an insomniac in a treatment used to reset biological rhythms.

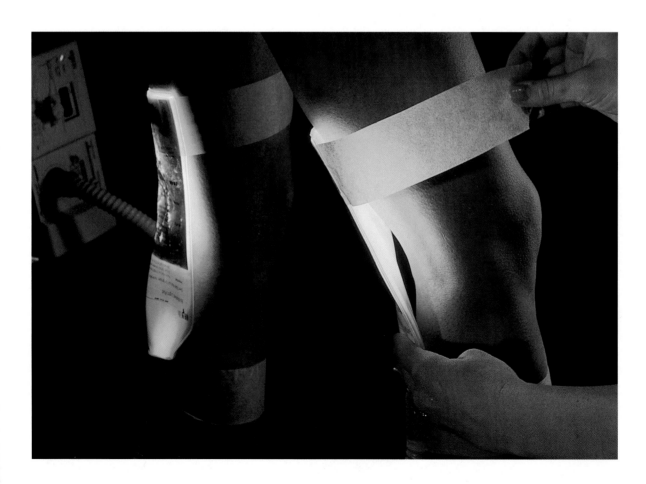

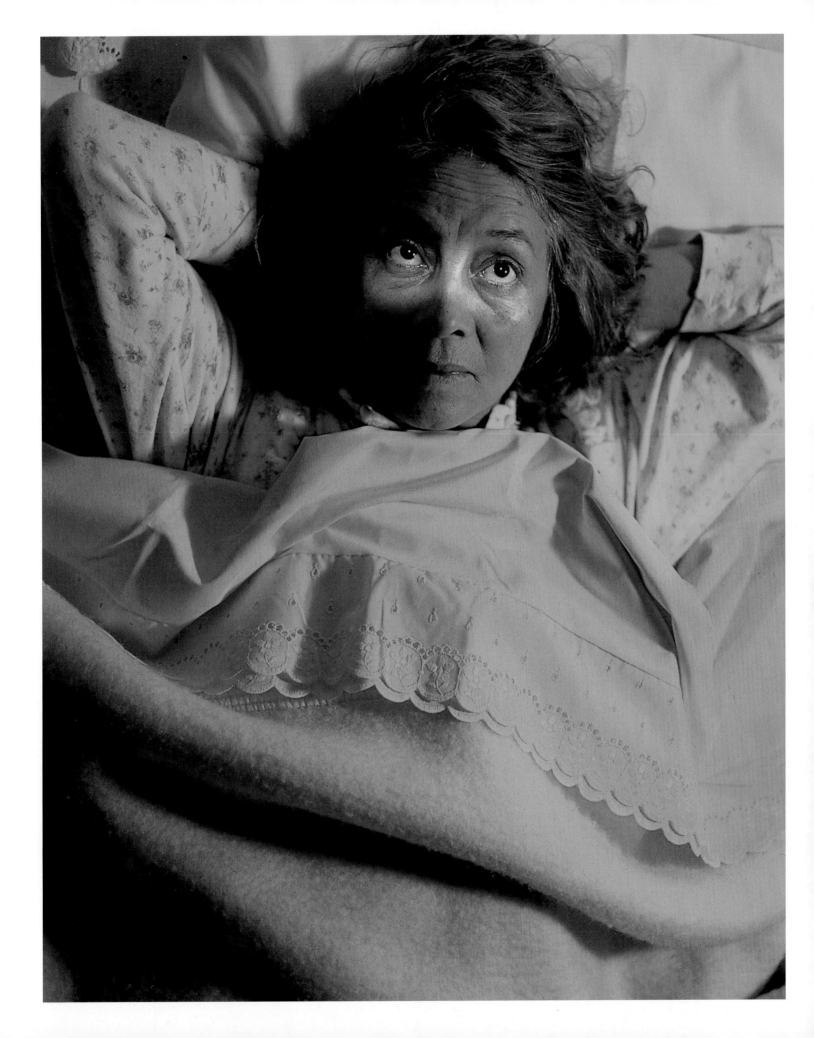

CHAPTER 2

TIMEKEEPERS OF THE ANCIENTS

> "The hunter knew only that
> the seasons change, the herds move on,
> and winter is nigh…. It was all one needed or
> could ever hope to know of time."

> — HISTORIAN GALE E. CHRISTIANSON

PRECEDING PAGES

**MEGALITHIC
MYSTERY:**

Astronomical calendar,
temple, or site of pagan
revels, the ancient
Stones of Stenness
stand mute on the
Orkney Islands.

**HEAVENLY
TIMEKEEPERS,**

Zodiac figures cavort
across a ceiling fresco at
the Palazzo Farnese in
Caprarola, Italy.
Imaginary trackers of
the sun's path, they
embody the time-
honored symbols of the
months, seasons, and
constellations.

JUST WHEN THE EPIPHANY occurred that enabled humans to replace a timeless present with a sense of measurable time will never be known, no more than we will ever know who first broiled meat over a fire or fashioned the first wheel. Remnants of early tools, fabricated to use in a forthcoming harvest or to kill the food for tomorrow's meal, are evidence of a sense of time, of planning for things to be done today and tomorrow. But it was not until human beings fully developed their most remarkable tools—language and numbers—that time began to be measured with accurate markers. And more than likely the first marker, the first recorded interval of time, was the day, the natural 24-hour period that multiplies into the broader seasons—seasons that can be tied to events in the natural world.

We cannot, of course, tell the time of day nor prepare a foolproof calendar by observing the movements of birds or the rising of floodwaters, no more than we can measure wind speed by a wreath of smoke. About all some of these events do is allow us to approximate. V-shaped flocks of Canada geese, for example, are harbingers of spring and winter in some regions, but, let's face it, their flights are not scheduled with the kind of predictability that we can base a calendar on. King salmon may swim upstream from the Pacific some time in July, but Americans don't schedule their celebration of Independence Day by such a trip. A monsoon (from the Arabic, *mawsim,* for "set time" or season) striking the Bay of Bengal reflects only a broad season, from April through October. And just because we plant beans when the gardener's manual tells us to doesn't mean we can fix a certain day in July when the crop will be ready for harvest.

Not all the rhythms of nature, impressive as they are, are useful when it comes to telling us the moment something will happen or how long it will last, when to be somewhere, how long to wait, what time to leave. If we are to make full use of time, we have to organize it better than even nature has. We must, as Jonathan Swift said, take it "by the forelock."

To tell the exact time of day—or identify a day, a month, or a year and place it on a calendar—we

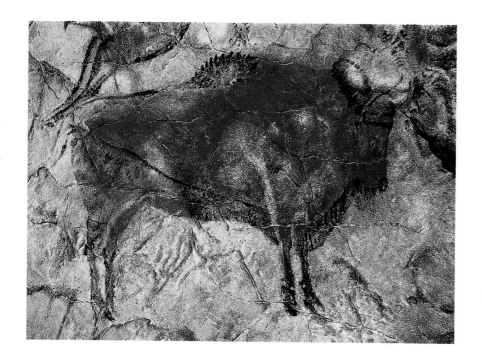

"ART IS LONG, AND TIME IS FLEETING," wrote poet Henry Wadsworth Longfellow. His words resonate in the primitive art that has endured as centuries have faded away. Beyond its beauty, such art also provides clear evidence of humankind's earliest awareness of time. The creative urge that immortalizes the present for future generations shows clearly in the lovely ivory head of a maiden, now 25,000 years old, recovered in France (right). The bone knife (left) found in an Ice Age cave in the Pyrenees may have served as an early record of seasonal change. Some 4,000 years ago, an unknown Paleolithic artist painted the bison (far left) discovered in a Spanish cave. The bisons' seasonal routines probably gave the people who hunted them, revered them, or both, an incipient recognition of annual cycles and therefore of time.

must identify cyclical natural events that recur, well, like clockwork. The most obvious such events are night following day, seasons following one another and piling up repetitively to form what we call centuries and millennia. Such concepts of long-term time were not always easily grasped. Although it may be difficult for us to conceive, very primitive folk probably lived in a state of perpetual and timeless present, like newborns and animals, confined by their world and unaware of either much of a past or a long-range future. "The individual possessed of the primitive consciousness of time" historian Gale E. Christianson wrote in *This Wild Abyss: The Story of the Men who Made Modern Astronomy*, "was circumscribed by the experiences of his own generation, and perhaps a few scattered reminders of his father's time. He had no written records to draw upon, none to leave, and precious few unique experiences to record, even if he had possessed knowledge of the written word."

WHILE EARLY HUMANS must have noticed the seasonal changes and the passage of day into night, they would have been hard-pressed, given their limited experience with the world, to know which events in nature—the gust of wind, a wet season, an Earth tremor—were relevant enough to time to be recorded. In any event, the exact length of a day or a year probably was not important to primitive peoples, who in the beginning needed only enough time to survive in the present. Later, their perceptions and needs would change as generations passed and experiences amassed. Powers of observation were honed, and humans slowly grasped the concept of cycles—not only in recurrent wintry snows and summer heat, in the migrations of animals, and the budding, blooming, and dormancy of trees, but also in the movements of the moon, the sun, and the stars and their relationship to the seasons. These

earlier humans must have gone from merely observing nature's rhythms to finding ways to record, measure, and use them.

Spiritual concepts quite naturally played an eventful role in human awareness of time, as various cultures sought to link in some way the obvious transitions that played out in the day and night sky, as well as the more illusive human transitions from birth to aging to death. Just as the moon, sun, and stars appeared to move unfailingly across the sky, so, too, must humans move, not physically but spiritually, through present time, perhaps toward some goal of endless time, marked by salvation or enlightenment. Virtually every religion conceived of time as associated with some cosmological event characterized by measured, constant, consistent change—a characterization that parallels the very process of timekeeping.

In religious and cultural practices, it's not difficult to find time-related concepts—rhythmic cosmological cycles, regularly occurring festivals, the belief in an afterlife, myths of renewal and rebirth—all ritualistic efforts to regenerate time itself again and again. For the Hindu, time is not a linear progression from past to future, as it is to the Westerner. Instead, it's a wide, cyclical swing through eternity, repeating itself like the seasons. The Egyptians worshiped the god of the midday sun—hawk-headed Re—who rose from chaos to generate a race that included the pharaohs and the gods Isis and Osiris, symbols of the renewal and continuity of life beyond time. For the Inca, time began when the sun god created the first Inca, Manco Capac, and his sister; they founded the Inca dynasty that continued for eight generations. The Shintoism of Japan—which venerates deities associated with natural forces and the emperor as the descendant of the sun goddess—sums up the theme of unrelenting recurrence with a proverb: "It is not good to hang next year's goddess of lucky direction too early." The "lucky direction" of

ART OR ARTIFACT:
Incised dots on this 15,000-year-old ivory plaque, excavated in Siberia, could be more than idle doodles. Like other undeciphered examples of ancient artwork, the spirals may reflect a pleasing design, a ritual symbol, or, just as likely, a record of time tracked through the movements of the sun, moon, and stars. Crude by modern standards, such an ancient calendar nonetheless could have told primitive people when to hunt, when to seek winter shelter, and when to anticipate the brief warmth of summer.

TIME TABLET:
In distinctive cuneiform script, a Babylonian tablet from 87 B.C. records the arrival of the comet now named after Edmund Halley. An English astronomer, Halley predicted that the comet, which made an appearance in 1682, would return in 1758. Records of virtually every one of its arrivals since 240 B.C. exist—solid evidence of the importance placed on the recurrence of natural events to track time. The time-sensitive Babylonians also kept precise records of the days between new moons, to produce the first calendar.

the deities is believed to change each year, so a new shelf for worshipping them must be made annually.

It was this concept of recurrence that gave rise to the calendar—in modern life typically an unimposing wall hanging but in reality an inspired, complex time tool. In general, calendars are based on religious responses to events in the cosmos and on careful astronomical observations, all having to do with the movement of the sun during the day and the motion of the moon and stars at night. In fact, our very word "calendar" is rooted in the movement of the moon and in the singular method used in ancient Rome to celebrate it. The Latin word *calare,* from which calendar is derived, means "to call or proclaim." This is exactly what ancient Roman priests did. Accompanied by ritual and the blare of trumpets, they called out the first day of each month, or *calend,* at the appearance of the new moon. Earlier than that, the megalithic monuments of Stonehenge, in present-day England, were erected beginning about 4000 B.C., apparently to keep track of lunar eclipses and solstices, as a way to track the seasons, planting time, and religious celebrations. Even farther back in time, some 30,000 years ago, Ice Age European hunters also saw value in the natural order of the moon's phases, scratching lines and punching holes in bones most probably as reference points for measuring the passage of the days.

But it was in Sumer some 5,000 years ago that the first real calendar was developed. The remarkable Sumerians, who occupied the lands between the Tigris and Euphrates Rivers, also created one of the world's first writing systems—cuneiform—and the first civil law code. The calendar they devised divided the year into 12 months of 30 days each. About eight centuries later, the Babylonians came up with a similar calendar of 360 days—a figure arrived at not because their astronomers weren't aware of the true number of days in a year, but because their system of

mathematics was based on the magical number 60, and many calculations, such as the 360 degrees in a circle, conformed to multiples of 60.

The Egyptians made up for the shortfall, adding five days to the year and setting them aside for feasting during the Nile's annual floods. They had discovered that the Dog Star—also called Sirius—rose next to the sun every 365 days, just when the flooding began. Probably created in 4236 B.C., the 365-day Egyptian calendar has given us what is presumably the earliest recorded year in history.

INDEPENDENTLY, OTHER CULTURES developed calendars of similar lengths, all relying in some way or other on the alternation of light and dark, crescent moon to crescent moon, the regular appearance of certain stars in constellations, and the recurrence of the seasons. Beginning some 2,500 years ago, the calendar masters of ancient Mesoamerica devised an intricate system that combined the 365-day solar cycle and a ritual cycle of 260 days, creating a grander cycle of 18,980 days—about 52 solar years. In a dazzling display of astronomical prowess, the Maya calculated the "year" of Venus at 584 days (it actually averages 583.92 days), then created mathematical tables that correlated the 365-day solar year with the planet's appearances as the evening and morning stars and its disappearance in between.

Like the Maya, the Aztec not only numbered the days of their ritual year but also gave them names such as "Water," "Flower," and "Death." Each combination produced varying degrees of good and bad fortune. The Aztec likewise divided their solar calendar into 18 months of 20 days each, with a period of five days, considered unlucky, at the end. At the culmination of the great cycle of 52 solar years, the Aztec performed human sacrifices in order to keep the universe in motion.

Not surprisingly, though, most early calendars

ENIGMATIC STONEHENGE: Poised to capture dawn's early light at midsummer, the giant megaliths of Stonehenge on England's Salisbury Plain have been reflecting the movements of heavenly bodies for some 4,000 years. Their origins have been variously attributed to Merlin, King Arthur's legendary wizard, and to the Druids, who purportedly used Stonehenge as a temple. The megaliths were unquestionably carefully aligned: The upright, horizontal, and concentric rings of stones suggest an observatory for predicting astronomical events. To this day, visitors gather at the ruins to watch the sun rise on June 21, the midsummer solstice and longest day of the year in the Northern Hemisphere.

were full of errors, usually due to reliance on the lunar month—new moon to new moon—as a convenient unit of time longer than the day. They defined the year based on the annual return of the seasons. These early, moonstruck civilizations calculated the moon's phases in 30-day cycles, which added up to only an approximation, five and a quarter days short of a true year. Since the moon's full cycle is actually 29.5 days, a lunar year works out to 354 days, 11.25 days shorter than the 365.25-day solar year. Astronomer Owen Gingerich at the Harvard-Smithsonian Center for Astrophysics, sorted it out, "Unfortunately, with a solar calendar, the number of 'moonths' in a year does not come out even. In the complex dance of the Earth and the Moon, the length of a month doesn't fit nicely into the length of the year."

ADJUSTMENTS BECAME NECESSARY, for if fractional errors were allowed to accumulate over time, the relationship of seasons to specific months would become totally skewed. A calendar that says it's July when it's still snowing outside isn't of much value to a farmer. To correct things, early mathematicians resorted to some deft juggling. Polynesians, for example, had a 13-month calendar regulated by the moon. "When the lunar year diverged too flagrantly from the procession of the seasons," historian Will Durant wrote of the way the Polynesians devised their calendar, "they dropped a moon and the balance was restored." Likewise, after Shang Dynasty astronomers in 14th-century B.C. in China developed a 12-month calendar, each month beginning with a new moon, they periodically added an extra month to bring the lunar record in accord with the seasons and the sun. The Babylonians eventually kept their months in some sort of sync with the moon by alternating 30-day months with months of 29 days, adding another 30-day month every so often to make up for days that had slipped by.

Early Greeks and Hebrews also inserted a month from time to time, the Greeks doing it three times in a cycle of eight years, the Hebrews seven times in a cycle of nineteen. Aristophanes's satirical comedy, *The Clouds,* written in 423 B.C., poked fun at the confusing aspects of the Greek calendar and its impact on festival dates. "When the gods were mourning, the Athenians were making merry;" Aristophanes wrote, "and when the Athenians ought to be sacrificing to the gods, their courts were in session, and their men were transacting routine business."

The earliest Roman calendars also suffered (or benefited, depending on one's expectations) from intercalating—inserting days to correct the difference between the lunar and solar years. Around the 8th century B.C., Numa Pompilius, the legendary second king of Rome, took on the current Roman calendar, a 10-month, 304-day contrivance attributed to one of Rome's alleged founders, Romulus. Reform in mind, Numa added January and February, inserting them between December and March. He also intercalated a month called Mercedinus, which alternated between 22 and 23 days and inserted it every second year between the 23rd and 24th of February. However, the Roman decemvirate, the ruling body of ten magistrates, added to the confusion by making other "corrections" that were in reality attempts to prolong their terms in office, hasten or delay elections, shorten the terms of their opponents, and make tax time come more frequently.

It was into this calendrical quicksand that Julius Caesar marched in 45 B.C., armed with knowledge of the Egyptian calendar and influenced by the Greco-Egyptian astronomer and mathematician, Sosigenes, author of *Revolving Spheres.* The Julian calendar was a purely solar one of 365 days but with an extra day added every fourth year in February. This solved the problem of the extra quarter day and gave us what we now refer to as leap year (so named

RAY CATCHER

TIME'S LONGEST DAY

The summer solstice is one of two moments in the year when the sun's apparent path, the ecliptic, is farthest north or south of the Earth's Equator. The word "solstice" derives from the Latin—*sol* for "sun", and *sistere*, "to stand still": At these times, the sun does seem to stand still in its northward or southward motion. In the Northern Hemisphere, the event occurs on June 21, making it the longest day. As the sun stops appearing higher in the sky, the days grow shorter, culminating on December 21 in the winter solstice, the shortest day of the year. After that, the days gradually begin to lengthen again, and the process begins anew. In the Southern Hemisphere, the situation, and the seasons, are reversed. Ancient farmers probably relied on the summer solstice in their hemisphere as a natural way to mark the changing seasons and to alert them when they should sow their crops. Since the summer solstice was associated with the growth and maturity of plants, it was also a time of celebration and ritual, often in recognition of fertility.

Arizona sunlight at the summer solstice works its way toward a bull's-eye—an ancient circle incised by humans to mark the moment.

because the extra day causes any date after February to "leap" over one day in the week and occur two days later than it did in the previous year, instead of the usual one). Caesar's manipulations—which gave us the same names, order, and length of the months we use in our current calendars—thus kept the calendar exactly in step with the seasons and with the time the Earth takes to complete its orbit.

But for all its foresightedness and order, the Julian calendar wasn't perfect. Because it was 11 minutes and 14 seconds longer than the solar year, the solstices and equinoxes drifted from their calendar dates. Not only were astronomical observations affected, but, more important to the Christian church, so were movable feast days like Easter, which was showing signs of slipping from a springtime holy day to a summer one. By A.D. 1582, the error compounded by Julius had increased to ten days, which meant that the civil year was ten days behind the solar year. Mathematical computations by the Bavarian Jesuit astronomer Christopher Clavius and a bold stroke of the pen by Pope Gregory XIII set things to right.

The Gregorian calendar (also known as the New Style calendar or the Christian calendar) was one of the pontificate's most spectacular achievements and continues to this day to guide the daily activities of most countries. As was true with many early calendars, Gregory's conception was a complicated, somewhat cumbersome process based on some deceivingly simple maneuvers. In a move as easy as tearing unwanted pages from a book, Gregory published a bull on March 1, 1582, striking 10 days from the Julian calendar simply by ordaining that October 5 of that year would be October 15. Moreover, to prevent further displacement, he ordered that, for centennial years, only those exactly divisible by 400 should be leap years. Put another way, three of the leap years that occur in 400 years would now be considered common calendar years *Continued on page 88*

Continued on page 88

Oldest and most basic of timekeepers, shadows are caused when the rays from a light source are cut off by an interposed opaque object. When that light source is the sun and it's high in the sky, shadows are short; when the sun is low, shadows are longer. Gnomons, the objects casting shadows, can be as monumental as the Pyramids of Giza (right), whose shadows move in a curve as the sun moves across the sky from east to west, dividing the day into broad swaths on each side of noon. A Syrian sandstone sundial (left) from the first century B.C. refined the measurements by subdividing time into smaller increments, shown by incised marks spaced evenly around the dial.

SUN AND WATER,
nature's nurturers, also
give substance to
featureless time. Deep
in an ancient well near
Aswan, Egypt, water
reflects the sun's rays
at the arrival of the
summer solstice. Using
calculations based on
the sun's midsummer
position almost directly
over Aswan, Greek
astronomer Eratosthenes
made a fairly accurate
estimation of the Earth's
circumference in the
third century B.C.
But long before that,
in 4235 B.C., the
Egyptians had become
the first people to devise
a calendar recording
the solar year.

PAYING HOMAGE TO THE SUN, an Egyptian papyrus from 1250 B.C. (below right) depicts a sacred scarab, symbol of resurrection, passing the sun to the creator god, Osiris. Like the sun, water's rising and falling also gave Egyptians an approximation of time. The Nile's annual floods, rendered in a Roman mosaic (far right), coincided with the appearance of the Dog Star, Sirius, every 365 days—a phenomenon that helped farmers calculate when to plant and harvest between the river's floods. On a smaller scale, an Egyptian water clock (bottom) marked time by the dropping of the water level as fluid leaked out slowly through a tiny hole in the clock's base.

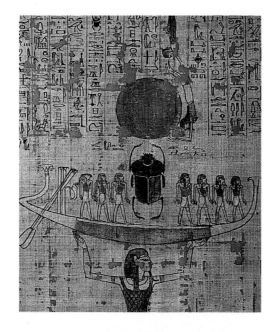

CALIBRATING AND CELEBRATING time have long been human passions. Among the ways to calibrate, calendars bring order to our lives, dividing the year into specific months, weeks, and days. But calendars do not exactly reflect the time Earth takes to complete its orbit around the sun: about, but not precisely, 365 days. To compensate for the inexactitude, Julius Caesar devised the so-called Julian calendar; the example at left is now preserved in an Italian museum. It added an extra day every fourth year, creating a leap year of 366 days. For centuries, the beginning of a new year has been heralded with celebration. The lavish 15th-century illuminated manuscript *Les Très Riches Heures* (right) portrays France's Charles, duc de Berry, presiding over a New Year's feast.

RESTRUCTURING TIME: In this 16th-century painting, Pope Gregory XIII presides over a conference aimed at correcting the Julian calendar. Ten days out of sync with the seasons, the Julian calendar was also adversely affecting the date of Christian feast days like Easter. The Gregorian calendar that resulted from the conference simply struck ten days from the year 1582, changing October 5 to October 15; in addition, the new calendar would no longer treat century years as leap years unless they were divisible by 400.

with no intercalary periods. The omission, in effect, of three days every 400 years meant that the error factor was reduced to only about two and a half days in 10,000 years.

By the end of the 16th century, most of Roman Catholic Europe had adopted the Gregorian calendar. The Protestants, who clung for a time to the Julian calendar because of their aversion to Roman dogma, eventually followed suit when merchants and lawmakers decided that reliance on two calendars—one of which was accumulating errors with every passing year—was an invitation to financial and societal disaster. Great Britain and its American colonies did not adopt the new calendar until 1752; by then, they were 11 days out of step with much of Europe. They fixed this simply by "suppressing" the third through the thirteenth of September.

EVEN THOUGH THE GREGORIAN calendar was both something new under the sun and something more in line with it, there were holdouts—and differences of opinion as to when a calendar should actually begin. The Gregorian calendar started with the supposed birth date of Jesus Christ (anno Domini, A.D., in the year of the Lord), a dating system conceived in 532 by a Scythian Christian monk, Dionysius Exiguus, and based on the tradition that Jesus was born "in the 28th year of the reign of Caesar Augustus." The exact date of Jesus' birth is unknown. The Christian fathers arrived at December 25 by first marking March 25 as the Feast of the Annunciation (when Jesus was supposedly conceived) and simply adding exactly nine months to that.

The Jewish calendar, on the other hand, begins with biblical creation, which is fixed at 3761 B.C. Based on the movement of the moon with the years corrected to solar time, it still keeps track of holy days and has a new year commencing with the month of Tishri, occurring at roughly the time of the Gregorian

calendar's September or October. Since the Jewish calendar is 11 days shorter than the solar year, intercalation necessarily intrudes in the form of a 13th month added 7 times during every 19-year cycle.

The 354-day, leap year-rich Muslim calendar, which relies exclusively on the motion of the moon, reckons that time began July 16, A.D. 622. That was the year of Muhammad's Hegira—his flight from Mecca to Medina. Used to mark religious feast days, the Islamic calendar requires a calculator and someone with a taste for math to make dates in it correspond to ones in the Gregorian calendar.

In 1793, the National Convention of the First French Republic introduced its own decimal calendar that divided the 12 months into decades and the days into 10 hours of 100 minutes each. Each month bore the name of its own botanical or agricultural characteristic—Germinal, Floreal, and Fructidor among them—nomenclature that was scrapped by Napoleon in 1806 in favor of the Gregorian calendar with its Julian-style months and names. Russia didn't adopt the Gregorian calendar until after the 1917 Bolshevik revolution, and, while some countries espousing the Eastern Orthodox faith have clung to the Julian rendition for the celebration of church feasts, Greece did adopt the Gregorian calendar in March 1924 for civil purposes.

But as workable as the Gregorian calendar is, even papal infallibility couldn't make it flawless. Months of unequal length must still be made to fit the solar year, a troublesome work of math that keeps dates and days of the week hopping through time. As a result, calendar reformists abound—among them proponents of the 364-day World calendar, a 12-month affair that assigns fixed dates to the days of the week and keeps all holidays on the same day each year. (To cope with the loss of a day, the calendar would add an extra day, Worldsday, between December and January, with another Worldsday

TIME FIX:

An 18th-century German perpetual calendar on a transportable card tracks the days of the week over a wide range of years, thus avoiding the need for a new calendar every year. Designed centuries ago, such calendars are not for the impatient: Instructions for one read: "To find the day of the week for any date, take the last two figures of the year date and add to them one quarter of the number formed by them, ignoring the rest."

inserted between June and July in leap years.) The World calendar's time has not yet come, probably because of opposition from religious conservatives. Other opponents of reform might be focused on something else: The World calendar assigns 31 days to April, a change that demotes our familiar, trustworthy—"Thirty days has September, April, June and November...." from valuable mnemonic device to inaccurate doggerel. There may also be a lack of interest because many of us simply feel that the familiar old Gregorian calendar hasn't broken yet. After all, a couple of skewed days every 10,000 years isn't worth a major overhaul when a few intercalations, or maybe some mathematical-calendrical legerdemain, can easily take care of the pileup.

WE TAKE FOR GRANTED our 24-hour day with its minutes and seconds and our calendars with their neatly laid-out boxes for the days of the month and the months of the year. But before such tidy timekeepers could be devised, the day's regular swaths needed to be divided and subdivided, broken down into smaller units of time just as the year is reduced to its monthly, weekly, and daily elements to make a calendar. The smaller units—which we know as seconds, minutes, and hours—are, like the week, artificial divisions of time, but are not, like months and years, based on the celestial phenomena of rotation and revolution. They came into being, as did the week, when humans realized they needed to impose a more exact regularity on life.

And so they dissected the day, using both the sun and the length and directional changes of the shadows it cast, as well as the nighttime appearance of certain stars and constellations. It was the Sumerians who, in addition to devising a calendar, first divided the day. They cut it into 12 periods (each the equivalent of two of our hours). These, in turn, were subdivided into 30 parts, each correspon-

ding to four of our minutes. Later, the Egyptians divided the day into 24 hours of two 12-hour cycles—generalized into 12 nighttime and 12 daytime hours to cover up the bothersome lengthening of daylight hours in summer.

Eventually, the time-telling astronomers had a language that would convert the regular alternation of day and night, the shadows, and the star patterns into a finely tuned system of definite time units that could actually measure the day. Credit for the system goes to the Babylonians, whose astronomers designed the sexagesimal system that we now use to affix numbers to the passage of time. Based on the number 60, which is easily divisible by 2, 3, 4, 5, 6, and 10, 20, and 30, their system is responsible for our convenient 60 minutes to the hour and 60 seconds to the minute.

Before numbers were allotted to time divisions, however, timekeepers made do with humankind's first clock, the shadow stick, or gnomon (from the Greek for "one that knows"). Early versions of this most ancient of time recorders were as basic as a tree that cast its shadow and as elaborate as the four-sided obelisks erected by Egyptians as long ago as 3500 B.C. Whether crude or refined, they all worked on the same principle: An object stuck in or placed on the ground and exposed to the sun will throw a shadow that moves steadily in a curve as the sun arcs across the sky from east to west. Observers realized that when the sun was high, the shadow was short, and that when it was low, the shadow was longer. At first, shadow clocks partitioned the day only into two broad segments on either side of noon and identified the year's longest and shortest days. They were used simply to tell general time—when to go into the fields or on a hunt, when to break for lunch, when to return from a hunt.

Eventually, sundials (dial comes from the Latin word for "day," *dies*) subdivided time more exactly through numbered markers arranged around the

FOLLOWING PAGES
RAVAGED BY TIME

but stately yet, the 900-year-old Temple of the Niches in Mexico's Veracruz state was built by the Huastec people as the centerpiece of their metropolis, El Tajin. The red glow created by this time explosure is reminiscent of the pyramid's original color. The 365 window-like niches lining its tiers may have been related to the 365-day solar cycle. Early calendar creators, the Huastec probably influenced the later Maya, who developed an elaborate and epoch-spanning calendar.

base of a monument or on a dial plane with a shadow-producing gnomon set parallel to the Earth's axis. Refined over the centuries, sundials even came in compact, collapsible pocket models with built-in compasses to orient the dial. One impressive hinged version made in 1636 had ivory halves inscribed with regular hours—Italian hours with a 24-hour day beginning at sunset, and Babylonian hours, with the day beginning at dawn.

For all their usefulness, sundials had several drawbacks, not the least of which was the one expressed in the familiar dial motto, "I mark only the sunny hours." Unfortunately, too, sundials were influenced by seasons, geographical location, and by the fact that solar time changes daily with the position of the sun; a variable shadow is, thus, produced, creating varying lengths for hours.

EFFORTS TO MAKE TIMEPIECES that were independent of the sun and still fulfilled the requirements of a clock—a repetitive, regular program marking equal increments of time—produced some novel substitutes. Burning candles was one. Since candles burn down at a steady rate, by embedding marbles or inserting pins to correspond with calibrated marks in the wax, one could, with some degree of accuracy, count off chunks of time as the candles melted to each mark. Similarly, the ancient Chinese burned knotted ropes, recording the time it took for the flame to move from one knot to the next. In a more elaborate Chinese variation, threads weighted on each end with a pair of metal balls were looped at carefully spaced intervals over a horizontal incense stick that was supported on a frame and set above a sounding tray. When the incense burned through a thread, its pair of balls dropped noisily onto the tray, signalling the passage of an increment of time.

Time's flow, as the ancients viewed it, could not have been given better substance, however, than in

two devices, the sandglass, or hourglass, and the water clock. An eighth-century glassblowing monk is sometimes attributed with the sandglass, still a familiar object in many households, notably in the three-minute version that cooks eggs to our liking. Simply a glass vessel with two compartments connected by a tiny aperture through which a measured quantity of sand, powdered rock, or crushed egg shell flows, sandglasses came in hour, half-hour, and half-minute models and were used for centuries to time a host of activities, from knightly jousts to the speed of ships to school hours to the length of pulpit sermons. Like all technology, however primitive, the sandglass had its drawbacks. While it was capable of catching what Samuel Johnson called "the transient hour," it was almost useless in damp weather, when the fine, uniformly sized sand particles clumped together and flowed more slowly through the aperture. Time literally dragged when a sandglass was malfunctioning.

The clepsydra, or water clock, also depended on the steady flow of a substantive material and not, like the sundial's shadow and the remains of a burnt candle, on a variable impression. Its name, derived from the Greek, means concealed or stolen water. (Historian Will Durant tells us of a Greek courtesan who was named Clepsydra, because she accepted and dismissed her lovers by the hourglass.) The principle is simple, but various methods from crude to quite sophisticated were devised to measure time intervals. Essentially, a water clock allows a graduated flow, or drip, of water from or into a reservoir through a small orifice. In its simplest form the clepsydra was a bowl with a hole in the bottom. Since the time it took to drain the vessel was always the same, one could generally calculate the hours just by observing the water level at marks on the inside of the bowl. A version of it was used in the Athenian and Roman courtrooms, *Continued on page 98*

TIME AND RELIGION

came together in the
Mesoamerican empire
of the Maya, who
worshiped the planet
Venus as a male god.
Using its orbital cycle
along with sophisticated
mathematics, they
devised an incredibly
accurate calendar.
Priests charted the heav-
ens from observatories
like this one at Chichen
Itza (opposite) on
Mexico's Yucatán
Peninsula; movements
of planets and other
heavenly bodies were
recorded and correlated
with the reigns of gods,
possibly with propitious
moments for human
sacrifices. In a Cacaxtlan
mural (right), a dancer
holding the five-pointed
half star, symbol of
Venus, may have
guarded a room where
captives were prepared
for sacrifice and put
to death.

FACE OF TIME:
Surrounded by a wealth of ritualistic and astronomical detail, the face of Aztec sun god Tonatiuh glares from the center of a massive calendar stone. A ritual cycle involving 20 days of the month and 18 months of the year, along with points of the compass, symbols of heaven, and even earlier world epochs, join with Aztec myth in the calendar to form a grand, encircling idea of the universe. Though artistically impressive, Aztec calendars were replete with errors, most due to a reliance on a 30-day lunar month.

where a speaker was allowed a set quantity of water for his speech, the quantity depending on the importance of his words.

Another version of the clepsydra sent water flowing at a uniform rate into a calibrated reservoir, where the rising level gave an approximation of the time. Introduced to Rome around 159 B.C., the clepsydra suffered from its tendency to freeze in winter—a quirk of nature confirmed by Julius Caesar. In his commentaries on his winter campaign in Britain, he noted that the freezing of water clocks threw his night watches into total disarray.

Without antifreeze in its tank, the water clock seemed doomed. But thanks to the ingenuity of the Egyptians and many other science-oriented experimenters, the clepsydra's design and accuracy were improved. Around the third century B.C., the Alexandrian physicist and inventor, Ctesibius—who also devised an organ operated by water and air, along with a force pump, a fire engine, and an air gun—added a series of gears driven by water to the clepsydra, turning it into a hydraulic, timekeeping wonder. Just as his organ gave musicians finer control over waves of sound, Ctesibius's water clock marked off the hours far more accurately than any other timekeeping device of the day.

Other kinds of timekeepers, the brainchildren of astronomers and astrologers as well as of horologists, kept track of lunar and planetary movements and, in so doing, told the time with bells and gongs or with tiny doors and windows that opened and closed as the hours passed. In the Athens of the first century B.C., the astronomer Andronicus of Cyrrhus constructed his Tower of the Winds, an octagonal, 42-foot-high building known in the Middle Ages as the Lantern of Demosthenes. Surmounted by a bronze weather vane in the shape of a Triton, it boasted a sundial on each of its eight faces, which were sculpted to represent the winds; and a 24-hour

SIFTING SANDS OF TIME

18th-century French sandglass

Similar in principle to a modern three-minute egg timer, the centuries-old sandglass can trickle away what Samuel Johnson called "the transient hour" and even break it into quarters. In this 18th-century French brass-and-glass model, each successive glass fills, then empties, to mark the passage of a quarter hour. When all the sand reaches the bottom, a full hour has passed and the device can be turned over to start timekeeping again. Aside from the skill that went into crafting them, sand-glasses were easy enough to create, since they relied on gravity rather than on tracking celestial movements. They could, however, slow time down if dampness kept their uniformly sized sands from flowing smoothly.

tion for more than three millennia, the Chinese contributed enough to mathematics, navigation, engineering, medicine, astronomy, and industry to help inspire the European industrial and agricultural revolutions. Quite naturally, innovations in mechanical clock design paralleled the Chinese interest in astronomy, and their first clocks perhaps only doubled as clocks; they were really symbolic, water-driven, astronomical instruments for observing moving heavenly bodies.

Waterpower continued to provide the driving force behind Chinese clocks, in part because the Chinese saw a connection between the flow of water and the movement of heavenly bodies. Su Sung, the creator of one extraordinary new clock, made the association quite clear in a letter to Emperor Zhe Zong in A.D. 1092, saying, "The principle for the use of water-power for the driving mechanism has always been the same. The heavens move without ceasing but so also does water flow. Thus, if the water is made to pour with perfect evenness then the comparison of the rotary movements (of the heavens and the machine) will show no discrepancy or contradiction, for the unresting follows the unceasing."

mechanized water clock. Gazing up at this edifice, marketplace visitors could tell the time of day and night, as well as get a fix on wind direction, the seasons, and astrological dates. Such elaborate, water-driven clockworks, monumental as they were, were a far cry from today's clock technology. Yet, in their gurgling, grinding, and grating movements was a strong hint of the mechanical clocks that would eventually provide humanity with a reliable means of measuring the abstraction of time, of synchronizing it, and of coordinating it with daily activities.

Not surprisingly, it was the Chinese who excelled in automating the water clock, thereby significantly narrowing the gap between ancient and modern timekeeping. Undisputed masters of inven-

OFTEN NATURE NEEDS an assist from technology. For the water clock to become more than a mere holding tank for water whose inflowing and outflowing told the time of day, more or less, some mechanism to distribute and use its energy more efficiently and exactly was required. That mechanism proved to be an escapement, a device that controls the motion of a clock's train of wheelworks, stopping it at set intervals and thereby enabling the clock to run steadily. The first clock to use such a mechanism was probably the "Water-Powered Celestial Map" (or, in a more literal translation, "Water-driven Spherical Bird's-Eye View Map of the Heavens"), built about A.D. 725 by I-Xing, a mathematician and Buddhist monk, and his engineer Liang Lingzan. His clock was an intricate bronze-and-iron array of wheels, shafts, interlocking rods and various stopping components, all housed in a wooden casing. The power that drove it did not come from a pendulum or falling weight, as in conventional 20th-century escapement clocks. It came from water flowing into scoops at the end of blades that turned a wheel, driving the rest of the clockwork.

As Su Sung described it, "Water was made to fall and rotate the driving-wheel automatically, so that the heavens, as represented, made one rotation in one day and one night. And there were two other rings (wheels) placed outside the representation of the heavens carrying images of the sun and moon, and also moving round. As the representation of the heavens moved westward one rotation…the sun and the moon met, and after 365 rotations the sun had completed one of its circuits…. Two wooden jacks (manikins), placed in front of the 'horizon' with a bell and drum, were made to strike the hours and quarters automatically."

SU SUNG'S OWN CLOCK took four years to build and was even more elaborate. Completed in 1092, it incorporated the water-driven escapement; stood more than 30 feet tall; had a waterwheel 11 feet in diameter; a bronze, power-driven armillary sphere (an ancient skeleton sphere ringed with metal bands representing all the imaginary circles of classical astronomy); and a rotating celestial globe. Three dozen scoops on the wheel caught water from a reservoir. When each cup was full, its weight would turn the wheel a notch, and the movement would be transmitted by gears to various time indicators. Also built into the structure was a five-story wooden pagoda with opening and closing doors that allowed admirers to see the wooden manikins sound the bells and drums and hold the tablets indicating the hour or special times of the day. "Thus," exulted Su Sung, justifiably proud of his creation, "we have a synthesis of the methods of the different schools."

But even such ingenuity wasn't enough to make the Chinese escapement clock perfect. While Su Sung's clock was a remarkable engineering accomplishment, water still froze in cold weather, an obstacle clockmakers of the day tried to overcome by placing burning torches nearby or by substituting mercury for the water. Also, the escapement was not actually automatic. A manually operated wheel raised the water in two stages from a sump beneath the drive wheel back to the reservoir above it, in a closed circuit system. The reservoir fed a constant-level tank from which water flowed at a regular pace into the drive wheel's scoops, one by one. Balance levers regulated the drive wheel and prevented a scoop from moving until it was completely full, dividing the wheel's rotation into equal units of time. But if human or mechanical error occurred, the regularity of the flow of water most likely would be disrupted—and time as told by the clock would be off.

Thus, water, for all its marvelous driving power, was a flawed energy source when it came to time-keeping. Because its rate of flow is so difficult to control accurately, any clock based on such a flow could never be accurate. It would take a future set of engineers to find another source of evenly measured oscillation. They discovered it in the force provided by a simple falling weight, and later by a swinging weight. Linked to more refined escapements, such gravitational force gave the world the first fully mechanical clocks—inspired devices that achieved their greatest glory in the West and provided more than merely glorious astronomical monuments. Centuries would pass before these clocks were transformed into the extraordinarily accurate timekeepers of today. Nonetheless, these early versions marked time as no man-made device ever had before.

GOING WITH THE FLOW OF TIME:
In an illumination from a 13th-century French Bible, Hezekiah, an Old Testament king of Judah, is guided by an elaborate clepsydra, or water clock. Like the sandglass, water clocks depended on the filling or emptying of a vessel at a steady rate. The Chaldeans of Babylonia may have invented it, but many cultures after them continued to elaborate on the concept. Galileo used one filled with mercury to time his experiments, and an oversized Roman version used a large cylinder into which water flowed from a reservoir; a float told the time as it moved up increments marked on the reservoir wall.

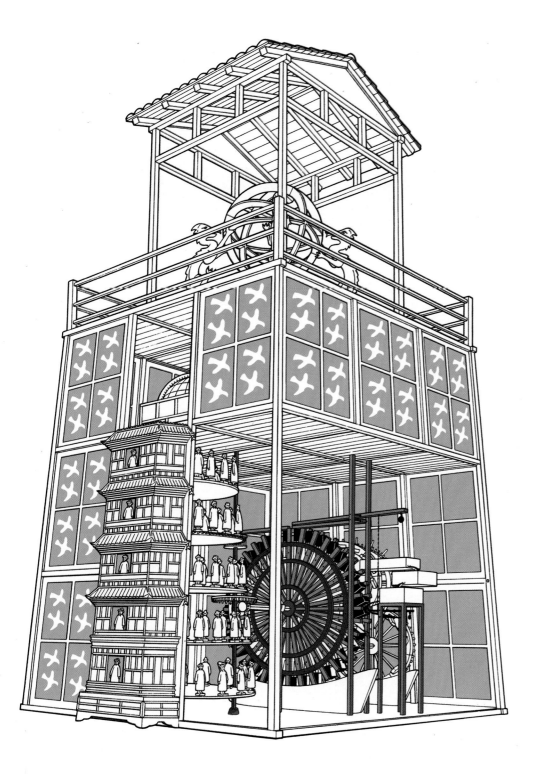

GRANDFATHER OF ALL CLOCKS, the great astronomical tower completed in 1092 as a memorial to Chinese Emperor Zhe Zong both observed the heavens and also heralded the age of automated timekeeping. Built by astronomer and inventor Su Sung, it stood more than 30 feet tall and was topped by an armillary sphere—bands representing the imaginary circles of the celestial sphere. A waterwheel 11 feet in diameter (re-creation at left) powered the extravaganza. As the wheel turned, shafts and chains moved various timekeeping parts, including manikins who rang bells and gongs to announce the hour. Controlling the water's flow and parceling it into equal intervals of movement—units of regular time—was the job of a water-driven escapement, like the working replica at right.

CHAPTER 3

MECHANIZING TIME

> "Time has no divisions to mark its passage, there is never a thunderstorm or blare of trumpets to announce the beginning of a new month or year. Even when a new century begins, it is only we mortals who ring the bells and fire off pistols."
>
> — THOMAS MANN

PRECEDING PAGES
PRAGUE'S FAMOUS TOWER CLOCK, crafted in the 14th century, tracks the sun's movements as well as the hours.

IN MEDIEVAL ENGLAND, inventive monks, like the one at top, produced clocks to chime the hours for prayer and worship. Centuries later, in Waterbury, Connecticut (left), Yankee clockmakers churned out thousands of alarm clocks that chimed Americans awake for another day on the job.

NO ONE REALLY KNOWS when the world's first nonwater-driven, truly mechanical clock came into existence, a clock regulated by machine movements uniform enough to give a close approximation of time. There is, however, some evidence that inventive monks built one around A.D. 1283 in the priory at Dunstable, a town northwest of London. That a monastery should be the birthplace of a clock is easily explained. The rules of monastic life dictated at least seven periods of worship each day. (Random prayer may have been equally good for the soul, but it carried no guarantee that the flock would always abide by it.) Thus, life in general was governed by carefully timed ritual. The Dunstable Priory clock, as horologists refer to that presumed first clock, was no doubt a rather crude affair. Like subsequent monastery clocks, it probably had no dial or hands and relied instead on bells that rang at calculated intervals, calling the faithful to prayer. (In fact, the word "clock" evolved from the medieval Latin *clocca* and the German word *Glocke,* both meaning "bell.") Falling weights, not water, produced the moving power for the first mechanical clocks. Hunks of lead or iron, they were attached to a rope slung around a drum. As they fell, the drum turned, activated an intermeshed gear train, and rang the bells.

Humans have found countless ways to regulate motion, whether by pressing an automobile's accelerator or adjusting the flow of water from a faucet. This first mechanical clock needed something to regulate and restrain the force exerted by the weights, just as the ancient escapement mechanisms in clepsydras controlled the energy released by water. The earliest such device was the verge and foliot, a clever contraption consisting of a pivoted spindle with projecting pallets (the verge) and a pivoted crossbar with adjustable weights (the foliot) that regulated the speed of the ticktock. Thus, the escapement was not only regulated by the rest of the clock but regulated the clock itself. Crude as it was, this new escapement would eventually transform the clock from a faceless, numberless system of alarm bells for cloistered monks to a household device with a bright, circular face around which the seconds, minutes, and hours ticked away.

Medieval clockmakers, even though restricted by relatively primitive technology and the high cost of constructing large timepieces, continually refined the shape of their verge-and-foliot escapements and improved their clocks' motive power.

By the mid-14th century, mechanical clocks with intricate gear trains, and sometimes even faces with one hand to indicate the nearest quarter hour, began to appear in the bell towers of great European cities. They were massive, elaborate, and, due to the widespread illiteracy of the time, usually without numbers. One of the first monumental mechanical clocks, the Parisian one set up by its German maker, Henry De Vick of Wurttemberg, in Charles V's palace tower (now the Louvre), was powered by a 500-pound weight that descended some 30 feet. A royal decree stipulated that its loud bell be used to synchronize all public clocks in Paris, putting an end to the daily cacophony of bells striking inaccurate hours.

In Italy a huge weight-driven astronomical clock built by Giovanni de Dondi, a professor of astronomy at the University of Padua, required an entire room to house its brass works. De Dondi's clock, 13 years in the making, was of heptagonal design, with dials showing 24-hour time, as well as lunar, solar, and planetary cycles; an engraved drum displayed the calendar day, the number of hours of daylight, and all movable feasts of the Catholic Church. Moved to a Spanish convent in 1556, the clock was destroyed by fire in 1809, but de Dondi's detailed descriptions allowed craftsmen to build several replicas and variations.

Some mechanical clocks entertained even as they informed. Such was the clock that graced the cathedral of Strasbourg beginning in 1350. Contemporary historian Daniel J. Boorstin described what happened when it struck the hour. "Three Magi bowed in procession before a statue of the Virgin Mary while a tune played on the carillon.... When rebuilt in 1574, the Strasbourg clock included a calendar showing movable feasts, a Copernican planetarium with revolutions of the planets, phases of the moon, eclipses.... Each of the four quarters of each hour was struck by a figure showing one of the Four Ages of Man: Infancy, Adolescence, Manhood, and Old Age."

Not to be outdone, other institutions created clocks of various sizes that vied with each other for richness and splendor of design. They might be fashioned of gold and embellished with diamonds, pearls, and rubies, or cast in the form of chariots, golden dogs, and birds with fluttering wings.

MECHANICAL ANGELS, gilded trumpeters heralding a procession of painted saints, and miniature jousting knights might have been a feast for medieval eyes, but did they tell accurate time? Barely, and certainly not by today's standards. Still, public clocks in high steeples, "sprinkling the quarters on the morning town," as Paris's urban architect, Baron Georges-Eugène Haussmann wrote, were imposing, a sign of the strength and wealth of a community. And they were a lot more palatable than the ear-splitting report of a noonday cannon or the nighttime bellowing of a town crier.

Despite all the horologists' fiddling with the foliot and the addition of crown wheels and striking trains, the first mechanical clocks all suffered from the same basic problem: the oscillation period of the verge escapement. The intervals that put the ticktock into time lacked the property known as isochronism, which refers to motions or vibrations of equal duration. There were reasons for the deficiency. First of all, the escapement in that era had no independent source of power and had to depend on the same system of weights that drove the clock. Also, clock parts were often tooled by gunsmiths, blacksmiths, or ironmongers, so the craftsmanship and the quality of materials were not always uniform. Finally, the escapements depended on the amount of driving force and were influenced by friction in the drive. Just as with the problem of water flowing into or out of a clepsydra, accuracy in these mechanical clocks was less than perfect, and they could be off by as much as two hours a day.

Still, despite their erratic movements, the verge escapements clanked on in their great iron frames for at least two centuries, mostly in turret clocks in town squares. A handful of scaled-down but still cumbersome domestic clocks were also conspicuously displayed in the homes of noblemen who could afford a one-handed decorative oddment that ran down within 24 hours. Eventually, inventors became increasingly aware that a good clock depended on a constant motion or vibration, and they hit upon a mechanism that would provide it: a metal mainspring coiled within a barrel.

Leonardo da Vinci first described this device in his notebooks but probably did not apply it to timekeeping. Still, in the early 1500s in Italy and southern Germany, clocks began appearing that replaced the heavy drive weights with a wound iron spring, which, when it uncoiled, turned an axle that sent the clock's pointer hand around in a clockwise circle. The invention could be installed in a small, drum-shaped clock—probably the world's first watch. The six-inch-high marvel was a European sensation, and soon watchmakers were competing for smallness, a feature sought after by those who appreciated portability at any price.

Soon, iron-cased miniature clocks and watches (the word "watch" probably stems from the Anglo-Saxon word, *wacian,* meaning "to wake," or "the night watch") were in widespread use. They were ideal now for shelves and tables and for watchmen, who wore them while making their rounds, either around their necks or suspended from a belt. These miniature clocks were, however, prone to malfunctions, especially when carried about.

ALTHOUGH THESE "NUREMBERG EGGS," as the small timepieces were called, told the time, they were still controlled by the problem-ridden verge-and-foliot escapement and thus were as inaccurate as their heavier, weight-driven cousins. The uncoiling spring provided adequate drive, but it was erratic: As the spring unraveled, its force varied, running too fast when it was first wound taut and slowing down after it had uncoiled somewhat. Constancy and regularity remained elusive concepts when it came to imbuing a man-made object with nature's precise, timekeeping rhythms. Still, 16th-century clockmakers were following the right path toward improved accuracy. The remarkable spring was still far better than cumbersome weights or running water. Devices such as the stackfreed and the fusee (the first a cam on which the spring pressed and the other a conical drum connected to the spring by a cord) equalized the spring's pull and helped make it a fairly reliable power source. But the verge escapement remained, regulating and telling time, just not necessarily the right time.

It was the great Italian scientist Galileo Galilei, contemporary of Shakespeare and Cervantes, who probably laid the foundation for accurate time measurement and paved the way for an improved escapement. In 1583, while sitting in the cathedral of Pisa, he reputedly was captivated by the to-and-fro movements of a great bronze lamp at the end of a long chain. As it moved repeatedly back and forth, he timed it with his pulse and observed that each swing of the lamp from one end of its arc to the other took the same amount of time, no matter the distance the lamp traveled. Galileo had discovered the pendulum. *Continued on page 116*

Continued on page 116

TOLLING BELLS,
like these in Sevilla's La Giralda tower (left), have long called the faithful to prayer. Indeed, the word "clock" derives form the Latin *clocca* and the German *Glocke,* both meaning "bell." The famed 19th-century painting "The Angelus," (right) by Jean-François Millet, depicts the bells' widespread effect: A peasant couple bow their heads in prayer at the ringing of the Angelus, a ritual repeated morning, noon, and evening in honor of the incarnation of Christ. In monasteries and abbeys throughout the world, bells still govern daily periods of worship, and, though they no longer serve as timekeepers for the rest of the world, church bells periodically interrupt the pace of modern life with their concerts of sweet, evocative sound.

GEARED FOR ACTION, the intricate winding mechanism of the 14th-century Wells Cathedral clock chimed the quarters and struck the hours in England until 1835. With a few new brass wheels replacing original iron ones, the clock's innards —which once tolled the hours with knights jousting and a figure known as a jack striking a bell—still tick away at the London Science Museum. Perhaps as early as 270 B.C. the Greek inventor Ctesibius of Alexandria had installed gear trains in clepsydras, or water clocks, but gearing may have been used in clocks even before that.

FAMILIAR FACES:
The clock face, with its time-telling dial, has become as recognizable as the face of a reliable friend, whether it is on a household clock or on a monumental public building. But, while clock faces may share similar features, their internal workings and the structures housing them take many shapes, from cumbersome clockworks and massive weights suspended inside clock towers to ingenious, open-sided domestic creations that display the timepiece's inner workings. This weight-driven, Swiss chamber clock (left) from 1572 has an interlocking frame made entirely of iron and a single-handed dial to indicate the hour and the nearest quarter. The town hall clock in the Basque city of Bilbao (opposite) has kept time for the town—and for revelers celebrating the annual festival of St. Fermin—for about three centuries.

Whether the lamp story is true or not isn't especially important. What is critical is that Galileo's work with gravity and motion included an interest in the pendulum. In his experiments, he tied weights to various lengths of cord and chain and observed their arcs of swing. Several facts emerged. For one thing, the heft of the weight didn't affect its speed from end to end of an arc. That is, a heavier weight didn't swing more quickly, nor did it matter whether the swing was great or small. The time it took for a weight to swing from one end of its arc to the other was always precisely the same, unless the length of the cord suspending the weight was varied. When the cord was lengthened or shortened, the period (or time duration) of the weight's swing was altered: The longer the cord, the more slowly the weight swung.

Galileo soon saw that a pendulum's movement over equal spaces in equal time made it an ideal measurer and regulator of time. In short, it possessed isochronism, the quality of equal timing essential to accurate clock mechanisms. Galileo also undoubtedly realized the potential for pendulum clocks, for he designed a mechanism that combined a pendulum with an escape wheel that kept the pendulum swinging as it controlled the escapement. But although both Galileo's son, Vincenzio, and his biographer Vincenzo Viviani had Galileo's instructions on how to build the clock, they apparently never got far beyond the sketch-and-model stage.

Still, the idea would not languish for long. In late 1656, 15 years after Galileo's death, the Dutch astronomer Christiaan Huygens, discoverer of Saturn's rings, designed the first clock regulated by a pendulum. A gravity-driven contrivance, it had a far more constant and natural period of oscillation than the foliot bar. Huygens's clock still needed a weight and a crown wheel for drive. But an improved escapement connected the pendulum directly to the verge that engaged the crown wheel, thereby efficiently transmitting the pendulum's rhythm to the clock. Impressively accurate, this clock had an error of less than a minute a day and was later improved to less than ten seconds.

Eighteen years after Huygens created his clock, he developed a balance wheel and spring assembly; the latter drew a plagiarism charge by English scientist Robert Hooke, who claimed priority. Still used in a few wristwatches today, Huygens's innovations allowed the clock to operate accurately in any position. Whereas in the older models the driving force actually controlled the clock, the pendulum was now the controller of the clock's driving force. Its advent was responsible for the tolling out of minutes and seconds as well as hours. The pendulum clock would remain the most precise instrument for measuring time well into the 20th century.

Even this ingenious, swinging mechanism would have its share of problems, often running, as the English poet Lord Byron said when he likened man to one, "betwixt a smile and tear." Gravity, latitude, and the weather's effects on the clock's various components conspired to alter the pendulum's swing. The weather especially had an adverse effect, even on a wooden pendulum. A rise in temperature lengthened the pendulum rod, causing it to swing more slowly in midsummer, while a temperature drop made the rod contract, thus speeding it up in winter. Again, refinements were not long in coming, and so-called compensation pendulums soon began to appear.

In the early 1700s, the English horologist George Graham substituted a mercury-filled vessel for the usual pendulum bob, the weight at its bottom end. When heat forced the pendulum rod to expand downward, the mercury expanded upward and counterbalanced the change. John Harrison, an English carpenter and self-taught clockmaker

ESCAPEMENTS
TIMEKEEPING BY TICK AND TOCK

Without an escapement device—and a means of regulating it, like a pendulum—a clock's movements would be aimless since there would be nothing to measure out what we perceive as time's procession of brief periods. A ratchet contrivance of varying design, an escapement is the mechanism that alternately checks and releases the gear train to tick out equal regular measures of time.

VERGE AND FOLIOT

The earliest clock regulator in common use, the verge-and-foliot escapement was invented around 1275 and used until the beginning of the 20th century. It consisted of a vertical shaft (the verge) with an attached horizontal bar balance (the foliot). The crown wheel teeth alternately engage a pair of pallets on the verge, causing the foliot, the escapement's regulator, to oscillate clockwise when the upper pallet is engaged and counterclockwise when the lower is engaged. Oscillation periods can be adjusted either by altering the weight powering the crown wheel or by shifting the small hanging weights on the foliot in or out.

ANCHOR RECOIL

Invented about 1671 and used in pendulum clocks, this escapement improved timekeeping to an accuracy of about one minute a week. It relied on the constant and natural period of oscillation of the clock's regulator, the pendulum. As the pendulum swings, it rocks the attached anchor, causing alternate pallets to catch and release the teeth on the escape wheel, whose regular forward motion is transmitted by a train of gears to the clock's hands. The escapement makes the wheels recoil slightly at every beat, giving the mechanism its name. It is often found in long case clocks, where the recoil of the second hand can be easily seen.

DEAD BEAT

In general use since 1715, the dead-beat escapement was actually invented some 40 years prior to that. Enhancing time-keeping accuracy to within a few seconds a month, it derived its name from the fact that the teeth of its escape wheel fell "dead" on the pallets with no recoil. Dead-beat pallets were often fitted with jeweled inserts to reduce wear and increase efficiency. Since there was less friction, timekeeping errors were greatly reduced. Clocks fitted with this escapement were fine-quality precision instruments often employed in astronomical observatories.

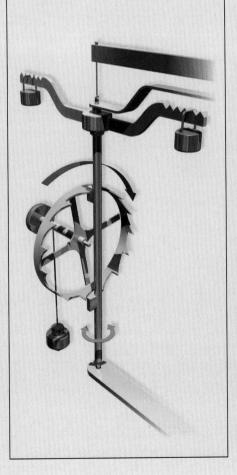

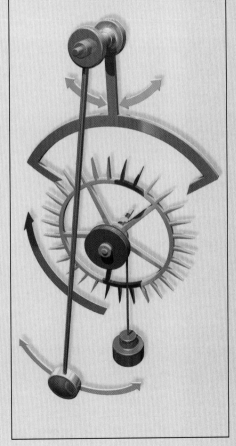

STEADY SWING:
Long a popular
attraction at the
National Museum of
American History in
Washington, D.C.,
a Foucault pendulum
toppled pegs in what
appeared to be a
clockwise rotation.
But the circular path
followed by the 240-
pound, gold-plated bob
was an illusion, since
it was the Earth that
moved beneath it, not
the pendulum itself.
Named for French
physicist J.B.L. Foucault,
who invented it in 1851,
the pendulum provided
the first laboratory
evidence that the Earth
rotates counterclockwise
on its axis. Two cen-
turies before Foucault,
Galileo had realized that
a pendulum's consistent
swing made it ideal for
measuring time.

(who, as we'll see later, solved another major time problem) refined the temperature-manipulating technique with a "gridiron" pendulum that reduced friction. This was an upright combination of brass, steel, and copper rods with different expansion coefficients; adjusting their lengths prevented any temperature change from affecting the pendulum's timekeeping property.

In the meantime, the escapements also got a makeover. A new anchor escapement was developed that improved upon Huygens's design by allowing a smaller pendulum swing arc and creating less interference to the pendulum's free movement. Later, Graham's "deadbeat" escapement came along, allowing the escape-wheel tooth to fall "dead" on the pallets without recoil, thus interfering even less with the pendulum's swing. These and other improvements made pendulum clocks far more regular than they had ever been, and by the end of the 18th century clock error had been reduced to only a few seconds a week.

Now also smaller, clocks were no longer just curiosities on the mantels of the wealthy or the walls of castles and country estates. Made at first by cottage industries, notably in the villages of Switzerland's Jura Mountains, clocks found their way into ordinary homes. In America mass production techniques developed after the Revolutionary War, making clocks even more accessible. Pendulum clocks still towered over great cities, chiming the time, but others now ticked out the hours, minutes, and seconds in all manner of compact enclosures.

The few early domestic clocks that had hung on walls or dangled from hooks and brackets were known as skeleton clocks, baring their inner workings through glass sides for all to see. Their replacements were aesthetically embellished, as well as functional. While "wag-on-the-wall" clocks—

early American clocks with uncased pendulums and weights—remained popular for a time, long-case, or grandfather clocks, and short-case grandmother clocks took advantage of the anchor escapement and narrow arc swing of the pendulum, as they enabled clockmakers to hang both weights and pendulums in slimmer, handcrafted wooden trunks. Cuckoo clocks with carved wooden birds that sang out the time emerged from Germany's Black Forest; "form" clocks took on the shapes of lanterns, birdcages, bowls of fruit, baskets of flowers, lighthouses, dogs, and banjos.

IN CONNECTICUT, home to some of the finest clockmakers in early America, Seth Thomas founded a company that bore his name and would remain one of the largest clock factories in the world until the 1930s. About the same time, another Connecticut clockmaker, Eli Terry, created America's first successfully mass-produced shelf clock, a "Terry-type" timepiece with delicate feet, slender pillars, a scrolled top, and works with wood or brass movements that had to be wound but once a day. The materials used to encase these new clocks' working parts were equally impressive: crystals to protect dials and hands; mahogany, rosewood, and ebony for the cases, along with occasional shell inlays; others were decorated with enamel mirrors, tortoiseshell, gold, and silver.

Pocket watches, too, evolved, and the original pomander-shaped portables invented around 1500 became obsolete. Many of the watches were now the circular shape that is standard to this day, but they were also sometimes made in the form of skulls, crosses, and miniature books; some had enameled scenes or paintings of historic figures on their hinged cases or were covered with precious metals, leather, horn, and sharkskin.

But decorative art and reduced size were not

GRANDFATHER CLOCKS, COMPACT WATCHES

The invention of the pendulum clock by Dutch scientist Christiaan Huygens was, until recently, one of the most significant advancements in precision time-telling. As clockmakers honed their craft, pendulums for long-case clocks were redesigned to overcome temperature effects, then shortened to fit into wall and table models. But it was the invention of the coiled spring and miniaturization of the gear trains that gave us the first dependable portable clocks and watches, like the ornate gold pocket timepiece at right. Eventually, mass production, interchangeable parts, and jeweled bearings put smaller, cheaper timepieces in homes and pockets. "The modern digital clock is more accurate certainly," says Dan Parkes (above), who restores antique clocks in his London workshop. "But if you're talking about the art of manufacture, something beautiful to look at, then it hasn't anything to offer."

PLEASING FORM AND INNOVATIVE FUNCTION have characterized clocks through the ages. In a Japanese silk painting, a wall clock from about 1700 gets a rewind from a kimono-clad woman (opposite, left). Practical as well as decorative, a clock set in an ormolu vase (opposite, right) came complete with automated singing birds and an organ at the base. Classically simple in design, an alarm clock (near right, above) and a stopwatch (far right, above) exemplify the necessity of precise timing in our lives. More complex, and a sign of modern technology's impact on time, a lemon-powered clock (right)—based on the original battery design of Alessandro Volta—runs on the electrolytes found in the fruit's acid juice.

the only refinements in store for the clocks and watches. More complicated and efficient inner works were devised, and, to underscore the value and necessity of time in so many human endeavors, there were new uses for the clock's self-regulating mechanism. By the late 17th century, doctors had "pulse watches" that could be stopped entirely, somewhat like a modern stopwatch. Clocks could track the moon and tell the time of high tide, while some pocket watches could wind automatically, using a weight activated by a person's movements; after the invention of the music box, also moved by clockworks, some even played a tune while keeping the time of day.

Mass production of clocks with interchangeable parts may have put a timepiece in more homes and pockets than ever before, but it was the art of jeweling that kept watches ticking longer and helped make timekeeping even more of a formalized, dependable, daily activity. Jeweling was born of necessity. As the coiled mainspring reduced the size of watches, their numerous rotating, sliding, and oscillating parts were in closer contact, and friction was increased. To alleviate the problem, bearings, which support and direct loads in moving machinery while reducing friction, were introduced in 1704 by Facio de Duillier, a Swiss horologist living in London. His tiny bearings were made of sapphires and rubies and were used as pivot holes because of their hardness and resistance to wear. Drilling holes small enough to carry the pivots required incredible precision, and the technique, at least in its early days, was shrouded in secrecy. Apprentices in Geneva who wished to learn more were asked to pay enormous fees—not only for the necessary tools but for the privilege of entering a rigidly controlled monopoly.

Jeweling remains an art to this day and can still determine the price of buying time, because the 7 to 23 jewels in a modern watch have their own intrinsic worth. However, synthetic forms of corundum, the hardest natural substance after diamonds, are also now used in watch bearings and have made watchmaking a less secretive and exclusive occupation—as well as making watches a cheaper and considerably more accurate commodity than they ever were before.

THE EARLY HOROLOGISTS' SEARCH for ways to make more accurate clocks was particularly important to people involved with the sea and shipping. The seafarers' interest was not merely academic, and it went far beyond an obvious need to schedule ship departures and arrivals. For one thing, the measured speed of a ship depends on time, just as the speed of anything that moves is recorded by time. Equally important, the determination of longitude depends, as we shall see, on accurate time telling: Without a precise way of assessing position, sailors simply could not tell where they were while out on a featureless sea. Time is the essence of all so-called STD (speed-time-distance) calculations. With simple equations and ordinary arithmetic, time is stamped into each of them: D equals ST, S equals D over T, and T equals D over S.

Today, the rate at which a vessel sails or is propelled by engines is figured by time-directed technology. Powerboats, for example, may have marine speedometers that rely on a small rotor mounted on the bottom of the boat, connected to a dial or digital indicator. A more likely instrument is a tachometer, which measures engine speed in revolutions per minute, information the pilot can translate into boat speed. Charts with logarithmic speed scales are another way of timing speed. Traditionally, navigators used a pair of dividers to estimate speed, but computers now make the process easier. No skipper today, whether operating

COMPASS AND SUNDIAL, this book-shaped, ivory tablet was used in 1599 to tell both time and direction. Compasses were used from the 12th century in Europe, and sundials at sea proved more reliable than clocks—but only when the sun shone.

FOLLOWING PAGES
PERSONIFYING THE TWELVE WINDS, puffed-cheeked figures adorn the ceiling of the Vatican's Tower of the Winds. Chartless sailors of yesteryear often followed coastlines to find their way—or they simply rode the winds.

in shallow or deep water, would consider a cruise of any duration without a reasonably accurate timepiece nearby. Errors in time accumulate over even a small portion of the day and can prevent one from determining how long it will take to get from here to there, not to mention fouling up other navigational reckoning as well.

In the early days of sailing, a vessel's speed could be calculated from a knowledge of distance and time, but a truly accurate timepiece, the crucial element in modern piloting, was not easily acquired. One time-honored way of determining speed at sea was called "heaving the log." A log line was attached to a thin wooden quadrant loaded with lead to keep it floating point up, and the whole affair was tossed into the water. Next, a half-minute sandglass known as a log glass was set, and, as the line paid out from a reel at the stern of the forward-moving ship, knots tied in the line at measured intervals were released, counted, and compared to the "sand" time passed. Each knot bore the same proportion to a mile that 30 seconds do to an hour. Thus, the number of knots that ran out determined the speed of the vessel, and the term knot came to represent one nautical mile an hour, a unit of sea speed equaling the internationally accepted measure of 6,076 feet or 1,852 meters per hour. Despite its failings—caused by fluctuations in tides, currents, winds, the rocking of vessels, the state of the sandglass, and the sometimes blurred eye of the mariners counting knots—the log served as a passable speedometer for centuries.

Instruments and techniques for assessing location at sea also had been long available, but at first only for determining latitude—the imaginary parallel lines circling the Earth in gradually shrinking rings from the Equator at 0° latitude to the Poles at 90° north and south respectively. Since the ancient Greeks, seafarers had realized that, as they traveled from north to south, the elevations of the sun and fixed stars shifted. The seasons too affected the look of the heavens. So, to find their latitude, the seafarers simply measured the height of the sun and stars over the horizon, using a variety of fairly sophisticated instruments that gave them a measurement accurate to within a fraction of a degree.

The astrolabe, described by the Alexandrian astronomer Ptolemy in the second century A.D. and refined by the Arabs, was a metal disk with a sighting device pivoted at its center. Engraved with star maps, circles showing equal altitude for given latitudes, and a circle of the zodiac identifying the sun's position throughout the year, it was suspended by a ring from the observer's thumb. The astrolabe not only measured the altitude of a celestial body but also converted the observation into an hour angle, allowing time to be determined by observing the sun. Used more by astronomers than seafarers in Europe during the Middle Ages (Columbus didn't use one), the astrolabe was eventually replaced by the sextant. This telescope-equipped instrument, invented in 1730, gave generations of sailors an easier and more reliable way to steer by the stars.

ASCERTAINING LONGITUDE at sea was not so simple for early sailors. Even an armchair admiral today can locate any place on Earth using the intersection of the parallels of latitude and the meridians of longitude that ring the Earth through the Poles north to south. But ancient mariners had to pour over maps that were artistic renderings more than they were scientific documents. Longitude lines were there, to be sure; they had been in place since Hipparchus, a celebrated Greek astronomer from the second century B.C., came up with a grid of 360 north-south-running lines that converged at the

"AND ALL I ASK IS A TALL SHIP and a star to steer her by," John Masefield wrote in his poem "Sea Fever." A woodcut from 1575 (left) captures such a moment at sea: navigators trying to locate the right star among a multitude to steer by. While the heavens gave direction, the seas provided the measure of a vessel's speed. At right, a 16th-century mariner holds a sandglass and log line to calculate speed. Attached to a float, the line was tossed into the water, while the sandglass was set for a half minute. As the line paid out from the stern of the forward-moving ship, knots on the line were released, counted, and compared to the time passed. Thus were the knots, or speed of the vessel, calculated.

Poles, along with 180 lines parallel to the Equator. Ptolemy also drew longitude and latitude lines on his famed atlas.

Anyone can draw lines on a map; what they mean is another matter. For where latitude lines, which measure angular distances north and south of the Equator are set naturally—the sun passing directly overhead at the Equator sees to that—longitude lines, which measure east-west angular distances, have no such naturally fixed starting point. They represent a distance east or west, but east or west of what? Sailors could easily tell where they were north and south of the Equator, but without any reliable 0° longitude from which to begin, they weren't always so sure about their location when they traveled between east and west. The open

sea remained as unmarked as nature had made it.

Ptolemy, for one, had tackled the problem of the starting point, choosing to run his 0° longitude, now known as a prime meridian, through the Madeira Islands off Africa's northwest coast. Seamen themselves were equally arbitrary, selecting their points of departure (or destination) as places from which to judge their east or west positions. In the meantime, mapmakers were increasingly using Ferro, an island in the Canaries, through which to run the prime meridian. But Paris, Rome, Copenhagen, and other cities would also be used before Greenwich, England, would at last be settled on as the site of the universal prime meridian.

Before Greenwich was agreed upon in 1884, however, the price paid for the shifting prime meridian was seagoing chaos. Magnetic compasses, which appeared in Western Europe in the 12th century and which some thought might be the key to finding longitude, would not be standard issue for some time. In any event, they were not all that accurate. Moreover, they were often regarded with some suspicion, and old hands preferred to rely on the informed guesswork known as dead reckoning—establishing position by a record of the courses sailed and the distances made on each course without celestial observations. Chartless sailors hugged the coast when they could. The charts they finally acquired were often little more than haphazard sailing directions based on even older charts and on oral histories of voyages. Some sailors simply rode the waves, the winds, and the currents, trusting to the forces of nature, God, and Lady Luck.

Understandably, crews sometimes lost their way and drowned in dangerous, uncharted waters. One notable incident in 1707 became a loud wake-up call demanding a solution to the seagoing chaos: A fleet of four British warships returning home miscalculated its whereabouts and plowed into the Isles

of Scilly, a smugglers' and pirates' haunt off Lands End, England's southwest point. Four ships were wrecked, and 2,000 men lost in one night. It was the worst nonaction disaster ever suffered by the Royal Navy. "In the case of the Scillies," the novelist John Fowles wrote, "the game was for long cruelly fixed against mariners. Before 1750, almost all charts, lazily copying ancient error, showed the islands ten miles north of their true position."

Whether due to divine intervention or blind luck, seamen in the age of sail beat all the odds on numerous occasions, managing somehow to explore new lands, wage war in distant waters, and engage in trade. What they lacked, of course, were the now commonplace instruments by which longitude is determined. First among these is an inexpensive pair of accurate shipboard clocks. One clock would be set to ship time, determined by the sun, and the other either to the time of the home port, to that of another site with a known longitude, or to the accepted prime meridian. By com-

paring the two times at exactly the same moment, a navigator could convert the time differences into a geographic coordinate to fix his position.

The determination can be made because the Earth revolves its full 360° in 24 hours, which means that one hour equals 15 degrees, or one twenty-fourth of a revolution. Tied to this rotation-dictated division is the difference in time at various places on the globe. As the Earth revolves on its axis from west to east, the local mean solar time at a location, which varies from location to location, advances by four minutes for every degree of longitude to the east. The farther one travels to the east, the later is the local time; the farther west, the earlier. Today, the world is divided into 24 time zones, under what we know as standard time, and the width of each zone is about 15 degrees of longitude. Thus, 15 degrees of longitude corresponds to exactly one hour of time difference. Time is, then, the measurer of longitude, which is itself expressed in degrees, with its subdivisions in the quite com-

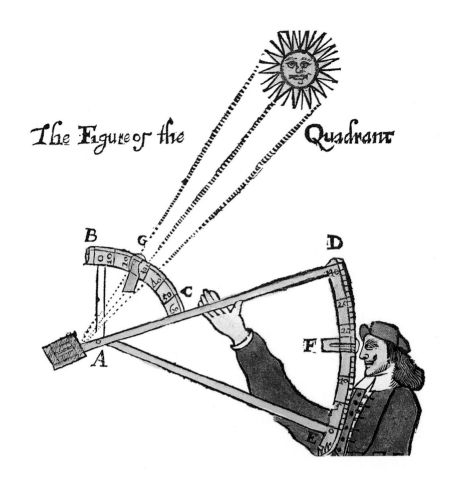

The Figure of the Quadrant

"SHOOTING" THE SUN with an unwieldy backstaff, an English mariner in a 1663 woodcut takes a sighting to determine the altitude of heavenly bodies. The instrument consisted of a graduated arc of 90 degrees and a reflector; its name, backstaff, derived from the fact that, when using it, observers would turn their backs to the sun. In 1730, both the backstaff and the astrolabe were superseded by the sextant. It measured the altitude of the sun or a star above the horizon; the angle of elevation and the exact time of day, registered by chronometer, made it possible to figure out the precise latitude with the help of published tables. While many early navigational instruments worked under calm conditions, mariners often had difficulty fixing them on a target star while standing on a heaving deck.

monplace clock markings of minutes and seconds.

To illustrate how it all works, let's say it's noon at the starting point, Greenwich, at 0° longitude. We know that when it is noon at a given location, it's noon at all other places on the same meridian, while it's forenoon to the west and afternoon to the east. When a ship arrives at a point 15° west of Greenwich, it is 11 a.m. by the onboard clock set to ship time; at 30° west it is 10 a.m.; at 45° west, 9 a.m., and so on.

Put another way, imagine that every day when a ship is at sea and the sun is directly overhead at noon, the navigator, using a sextant, accurately determines the time for his meridian of longitude by taking a "sight" of the sun or stars. He then resets his onboard clock to that "local" noon and compares that time to his "home" time. He knows that he'll pick up 15 degrees of longitude with every hourly shift up to a maximum of 180 degrees in each direction, east and west of Greenwich. Moreover, the 15 degrees of longitude translate into

distance. Unlike the parallels of latitude, which are evenly spaced with the distance between each the same (one degree of latitude is around 60 nautical miles), the meridians of longitude are widest at the Equator, but they draw closer together as they run toward the poles. Thus, while one degree of longitude equals four minutes of time the world over, the distance between adjacent meridians varies with latitude. At the Equator approximately one degree equals 69 land miles or around a thousand miles every 15 degrees; at 60° north or south, a degree is worth around 35 miles; at the poles, it's zero in terms of distance.

Simple as it all is to grasp now, finding longitude and nailing down its relationship to time took centuries. It taxed the imaginations and all but exhausted the expertise of a phalanx of celebrated scientists, navigators, kings, and astronomers—from Galileo to Charles II of England, from Newton to Philip III of Spain.

Galileo's contributions to the project had less

to do with any experience at sea than with his abiding interest in time as a curiosity of natural philosophy and as a tool in numerous experiments. His studies of motion, in which he rolled balls down inclined planes, relied on a water clock of his own design. "For the measurement of time," he says of that research, "we employed a large vessel of water placed in an elevated position; to the bottom of this vessel was soldered a pipe of small diameter giving a thin jet of water, which we collected in a small glass.... The water thus collected was weighed, after each descent, on a very accurate balance."

There was also his work in establishing the isochronicity of pendulums. But as Gilberto Bernardini, director of Scuola Normale of Pisa, has observed, "There is something more: Galileo introduced time as a normal coordinate. In his diagrams he drew a straight line as the time axis on which he marked time intervals from a zero point chosen at will. Today, everybody carries a watch, and reads time-space diagrams. But three centuries ago, time was only the correlation between the cyclic conception of day and night, of moons and years, and the unresting evolution that brings everything but human souls to an end."

MOONS CERTAINLY were part of Galileo's proposed solution to the longitude problem. When Philip III offered a pension to the man who could discover the elusive mystery of longitude, Galileo informed him that the answer lay in the heavens, but not in the conventional time-measuring stars, moon, or sun. Rather, it lay in a universal clock composed of the four moons of Jupiter, which Galileo himself had recently discovered. He noted that eclipses of the moons, brought about when the satellites entered the shadow behind the planet, occurred like clockwork, many times annually. The phenomenon, he suggested, could make of

the moons a universal clock. By tabulating the time of each moon's expected disappearance and its reappearance over a fixed point of land and then comparing that information to the time the eclipses occurred over a ship at sea, the result would be similar, at least theoretically, to what occurs when local time is compared to Greenwich time: A time difference converted to a difference in longitude, with an hour equal to 15 degrees. He even devised the celatone, a gas mask-like helmet pierced by a telescope, to make the necessary observations.

Although Galileo's reasoning was sound, his plan was impractical. He had observed his moons mostly from stable land, but a navigator on the rocking deck of a ship would have a difficult enough time keeping his own balance, let alone perusing complex eclipse tables and tracking Jovian satellites with a hand-held telescope of questionable power, or worse with a celatone strapped on his head. So the incompatible shipboard environment, coupled with Jupiter's invisibility during daylight and even at night for portions of the year, made the idea impractical.

Galileo's work, however, was not wasted. A dozen or so years after his death in 1642, French mapmakers and surveyors saw his moon clocks as a viable way to chart the land if not the seas. Spurred on by Gian Domenico Cassini, the Italian-born French astronomer who had constructed impressively accurate tables of the motion of the Jovian satellites, the French cartographers used the tables to revamp topography and came up with a new map of France that sliced degrees off the west and Mediterranean coasts. Viewing the finished work, King Louis XIV is reported to have remarked that his astronomers were taking more of his territory than were his enemies.

In the meantime, Cassini was directing a newly constructed national observatory in Paris. John

Flamsteed, a clergyman who became England's first Astronomer Royal, was in charge of a similar facility in Greenwich, with the same aim: celestial observations that would enable navigators to calculate longitude accurately. Longitude was time, and for the shipping industry the two elusive concepts could be used to expand commerce.

But how to put it all together? Lunar observations seemed to be the key elements in the effort, since the moon's movement across the sky was relatively rapid in relation to fixed stars. By measuring the angular distance between the moon—which would serve as a universal clock—and a selected star, time "at home" could be set and compared to the "local" time obtained at sea by observing the sun. The difference, as in Galileo's method, would produce longitude.

Still, the lunar distance method for finding longitude needed work to be successful. Labor intensive, it was, again, difficult to do aboard a rocking ship, and it required improved instruments and precise tables. Eventually, it became clear that an easier way to find longitude was to carry a clock aboard set to Greenwich (prime meridian) time. Then one could compare the moment of, say, sunrise with the time shown at that very instant on the objective clock. Simple enough, but if the clock was inaccurate, the calculation could be dangerously off. And the degree of accuracy that was required—especially if the clock were a pendulum model that had to be carried aboard a pitching, rolling ship—was not something easily achieved. Even Isaac Newton, who touted the value of a sea clock, had to admit that such a timepiece would be a long time in coming due to a ship's motion, variations in temperature, wet periods, and gravity differences in certain latitudes.

In 1714, probably spurred by the Isles of Scilly disaster, the British government convened a board of longitude whose aim was to produce a longitude-determining clock. As an inducement, the board offered a cash prize worth more than two million dollars in today's money to anyone who could construct a clock that would give longitude to within half a degree. John Harrison, the self-taught clockmaker, was among those who took up the challenge. His first chronometer was no mantel doodad. Built in 1734, it weighed in at more than 70 pounds, but it was constructed to take the punishment it would surely encounter at sea. The mechanism was essentially the same as that in a common watch today.

Almost friction free and made of rustproof materials, Harrison's chronometer was driven by a spring, the power from which was cleverly regulated so that it remained uniform at all times, no matter how much the vessel heaved. Changes in temperature, which had rendered other spring-powered watches inaccurate, were compensated for with a balance made of metals of different coefficients of expansion: When one part of the clock expanded or contracted, another did the opposite, thus varying spring tension and ensuring a constant beat of time. To further ensure that the ship's motion did not affect the clock's accuracy, it was hung in a frame of balancing rings, called gimbals, which allowed it to remain horizontal whatever the ship's position.

Preliminary sea trials were encouraging, but, even though the measurement that had plagued navigators for centuries was on the verge of solution, Harrison would have to wait for some time to collect his prize. For one thing, the board of longitude was made up of astronomers, scientists, government officials, and naval officers, who were still committed to the lunar system of celestial mechanics. Put another way, one could say the board leaned toward astronomers, not toward a

COSTLY MISCALCULATION: Sailing into uncharted waters without a precise way of determining longitude, mariners often relied on outdated charts and oral histories. The consequence: countless shipwrecks. Here, a Spanish galleon runs aground on a Pacific reef off Saipan. It took 18th-century English clockmaker John Harrison to come up with a solution to such miscues: Several versions of his sturdy, seagoing chronometer (right) calculated longitude with remarkable accuracy.

carpenter's son who had managed somehow to come up with the navigation breakthrough of the century. The first chronometer was also too expensive, too big, and it hadn't fulfilled the stipulation of half-degree accuracy.

Harrison refined and scaled down his creation, building three more chronometers over the next 25 years, the last about the size of a large watch. In 1761, Harrison's "Number 4" sailed aboard a frigate bound for Jamaica. The device performed incredibly well, keeping time to about a fifth of a second a day, almost as well as a land-based pendulum clock. When the ship reached Jamaica, the chronometer's longitude calculation, aided by properly timed observations of the sun's altitude, was accurate to one-fiftieth of a degree.

Capt. James Cook took a copy of a Harrison chronometer on his first *Resolution* voyage toward the South Pole. The clock lost only a bit over two minutes a year. It was "our trusty friend, the Watch," Cook exulted, "our never-failing guide." Harrison eventually got his prize money but only after petitioning George III for it. Although his chronometer continued to behave superbly, there was, as had been true throughout the history of clockmaking, room for improvement. Pierre LeRoy, a French scientist working on his own, moved several steps ahead of Harrison, when in 1765 he designed a more accurate longitude clock, one that contained all the elements of the modern chronometer. Central to his invention were an automatic system that adjusted for temperature and an escapement that was separated from the drive mechanism, thus reducing friction even more. Longitude, once as elusive as time itself, no longer bedeviled navigators and clockmakers. At last, it was a clearly defined component of a global grid system, one that could be measured as easily as the hours, minutes, and seconds that shape it.

Though the chronometer had solved the longitude problem, it had not fixed the location of a universal time base. For some time, mapmakers counted longitude west from a variety of locations: the Naval Observatory in Washington, D.C., the Paris Observatory, and the Royal Greenwich Observatory. In 1884 a conference was convened in Washington, D.C., to decide the matter, and delegates from 26 of the world's major countries attended. The U.S. and Britain favored Greenwich (Jerusalem and the Great Pyramid were also among the contenders), but the French held out for Paris. Though Greenwich was eventually accepted, the French insisted for a time on drawing their maps to the prime meridian at Paris.

England's leadership role in exploration and in producing accurate navigational maps was perhaps a deciding factor in the selection of Greenwich, but so, too, was the International Date Line, the place where each calendar day begins. When travelers heading west cross the date line, they must change time by setting their calendars ahead 24 hours; traveling east, calendars are set back a day. If the date line had run through a large country, setting calendar dates and time-keeping in general would have been skewed for that country. But because the Greenwich meridian is set at zero, the date line veers easily around political boundaries and more or less through the middle of the Pacific Ocean, hitting only small patches of land.

Greenwich thus became not only longitude's home port but time's hometown, the place where the global time zones began. The little town on the Thames was the answer to a modern society's need to avoid time discrepancies, the spot where the civil time we now know as standard time begins. Time would never again be "the same," as we'll see in the next chapter.

EARTH'S GRAND GRID of imaginary latitude and longitude lines supports the giant mapparium at the Christian Science Publishing Society's Boston headquarters. Thirty feet in diameter and made to be viewed from inside, the glowing globe portrays the world's boundaries in the early 1930s. Each of the 608 stained glass panels covers 10 degrees of longitude and latitude. Here a workman crosses time zones with ease as he polishes wide swaths of the Earth.

CHAPTER 4

FINE-TUNING TIME

> "We still cannot say what time is;
> we cannot agree whether there is one time or many
> times, cannot even agree whether time is an essential
> ingredient of the universe or whether it is
> the grand illusion of the human intellect."
>
> — PHILIP J. DAVIS AND REUBEN HERSH

PRECEDING PAGES GLOWING AND CONFIDENT, the big four-face clock in New York's Grand Central Station has served as a beacon for commuters coming and going since 1913.

CRITICAL TIME: Computer-assisted emergency dispatch systems like the one in Largo, Florida, help cut vital response times for ambulances. Medical emergencies are among a host of situations dependent upon precise timing.

BY THE MID-19TH CENTURY, highly accurate marine chronometers were ticking off the time on the high seas from the Antarctic to the Equator to the Arctic Circle. But landlubbers who didn't travel far outside their own communities and rarely if ever crossed longitude lines continued to live by standardized time intervals —seconds, minutes, and hours—rather than by a standard time. Residents of a given region made do with time as determined by the localized moments marked by sunrise, noon, and sunset. Which meant that every town and village kept its own time, and anyone with a watch in his or her pocket or a clock on the mantelpiece could make it to work or keep a local appointment. The general population probably thought, if they thought about it at all, that everyone else in the country was following the same schedule as theirs at the same time.

Of course, synchronizing local time required very little effort: All one had to do was observe the sun directly overhead at high noon, listen for the blast of a factory or a mine steam whistle, or watch a "time ball" drop at solar noon from the top of a pole placed so as to be visible at some distance. But the problem, as most of America eventually discovered, is that sun time is not the same time across the continent. The solar day, as measured from sunrise to sunset, varies in length from place to place, depending on latitude, and the time when the sun is overhead also shifts as one travels longitudinally around the Earth.

No American before the Civil War seemed too concerned about such astronomical vagaries, and coordinating local time with apparent solar movement worked nicely—as long as high-speed transportation was a futuristic dream. Change, however, was not only in the wind but already existed overhead. As villagers naively set their clocks by their local sun time, that time was shifting quietly about a minute for every dozen miles east or west. Which meant that there were really hundreds of local times, each different. Again, this did not pose much of a problem for a leisurely ride through the countryside, but it grew increasingly bothersome as stagecoaches—and later one of time's greatest arbiters, the railroads—became relatively speedy, long-distance carriers, faced with scheduling many different departures and arrivals.

As travel time shortened, life for the eager new travelers brought a mixed blessing. They might get to a destination faster than ever before, but if they went by rail or stagecoach, the ride there was a rocky, and more often than not, untimely one. The sun was still running the show, but at its own pace and in its own time, and that meant that local time was coordinating local affairs. Moreover, local time rarely jibed with that on train-crew watches, which were being set by the local time at their main terminal. With railroad lines now fanning out across the country, the roadbeds ran through cities and towns that kept their own time of day and could hardly care less about what the time might be at some distant railhead. Anyway, hours were still what counted when it came down to local lifestyles—the clear intervals of time that signaled wake up, breakfast, lunch, dinner, job shifts, and bedding down for the night.

RAILROADS WEREN'T THE ONLY businesses worried about making time more precise. With industrialism had come inventions that produced more goods faster—steam engines, spinning jennies, power looms—and getting products and the people who made them to the right place at the right time was essential to success. All of time's increments were now money. Without correct timing, opportunity was lost. Schedules, connections, switching were topmost in the minds of rail operators. A few minutes here or there may not make much difference to people going about their daily business, but to a railroad running fast trains, minutes could mean missed trips, late or haphazard arrivals, even wrecks.

In an effort to regulate ever tighter train schedules, the railroads eventually posted timetables, called "arrangements of trains," in their stations and tried to stick to them. But there was still a major obstacle: Those versions of "local" time rode the rails out of different terminals. "Local" time wasn't "local" anymore when the train arrived somewhere else. Thus, for all their attentiveness, the railroads had actually made matters worse. The railroad station in the city of Buffalo stood as one notable example of the confusion all this generated. It boasted three large clocks, one set to Buffalo time, another to New York City time (favored by the New York Central Railroad), and the third to time in Columbus, Ohio (favored by the Michigan Southern). Stewart H. Holbrook, a prolific writer and one of America's noted informal historians, described the effects such stubborn adherence to local time had on the railroads in the 1800s: "The Pennsylvania used Philadelphia time in the East, which was five minutes slower than New York City's, but five minutes faster than Baltimore time. The B&O system was very complicated. It used Baltimore time for trains running out of Baltimore, Columbus time for trains in Ohio, Vincennes time for trains west of Cincinnati....There were at least 27 local times in Michigan, 38 in Wisconsin, 27 in Illinois, and 23 in Indiana. A careful traveler, going from Eastport, Maine, to San Francisco changed his stem-winder 20 times during the trip."

With passengers, shippers, engineers, and conductors growing more bewildered and irritated, something had to be done before inconvenience exploded into total chaos. Working on the railroad just to pass the time of day was fine as a song lyric, but it didn't do much for smooth operation. There was, however, a glimmer of light at the end of the tunnel. Charles Ferdinand Dowd, the headmaster of a seminary for young women at Saratoga Springs, New York, started firing off letters and pamphlets to railroad executives. He urged them to adopt a system of national railroad time, one that would divide the continent into time belts, each an hour wide and within any one of which time would be standardized. The response was overwhelmingly positive, but in the ensuing financial panic of 1873, the plan was shelved. Its time would come, however, for a group of railroad officials had formed a permanent organization known variously as the General Time Convention or the Time-table Convention, which became the Association of American Railroads. They resurrected Dowd's suggestions and selected William Frederick Allen, a former engineer and managing editor of the influential *Official Guide of the Railways and Steam Navigation Lines,* to refine them.

Allen rejected the idea of using meridians an even hour apart and instead based his zones on the zigzags of geography, on economics, cities, and the general habits of the populace. His plan, as adopted on October 11, 1883, by the General Time Convention, split the 48 contiguous states into four standard time zones, each based on a meridian west of Greenwich: the 75th, 90th, 105th, and 120th. And so the country was divided into now familiar eastern standard, central standard, mountain standard, and Pacific standard times. The next year, at the convention that established Greenwich as the prime meridian, the world followed suit and divided itself into 24 time zones, each 15 degrees longitude wide, an hour apart, divided east and west of the zero-degree prime to the International Date Line.

A standardized time had now been imposed on the world, and the political and geographical considerations that forced some of the international time zones to break the 15-degree rule or delineate areas that differed by half an hour were not apparently issues of concern for many people. In 1883 it was announced that at noon Sunday, November 18, Allen's time zone plan would take affect. It would be set in motion by Greenwich mean time being broadcast over telegraph lines to major cities, where authorities would adjust their clocks to their zone's proper time. But not all Americans were ready to comply. As historian Holbrook tells it, preachers who abhorred the railroads because they ran on the Sabbath urged their flocks to continue their Christian ways on Christian time; some characterized the whole idea of standard time as a lie that should be ignored. Holbrook reported that the mayor of Bangor, Maine, "…vetoed a city ordinance that provided Bangor should adopt Eastern Standard Time, and with great feeling, shouted, 'It is unconstitutional, being an attempt to change the immutable laws of God Almighty and hard on the workingman by changing day into night.' He even went further and threatened to have his constables prevent the sextons from ringing the church bells on the new and unspeakable time."

In Indiana, where some cities also delayed adopting the new standard, the *Indianapolis Sentinel* was almost hysterical: "The sun is no longer boss of the people. People, 55,000,000 of them, must eat, sleep, and work as well as travel by railroad time. It is a revolt, a rebellion. The sun will be requested to rise and set by railroad time. The planets must, in the future, make their circuits by such timetables as railroad magnates arrange.… People will have to marry by railroad time, die by railroad time."

Despite such scattered protests, the change to standard time was a welcome relief, especially for the railroads. Interestingly, though, the federal government wasn't ready for standard time, arguing through the attorney general that no government department could adopt it until authorized by Congress. "The railroads went right ahead with the plan," Holbrook explained, "and the Attorney General, according to a good and perhaps

apocryphal story, went to the Washington depot late in the afternoon of the 18th of November to take a train for Philadelphia. He was greatly astonished…to find he was exactly eight minutes and 20 seconds too late." Still, it was 1918 before lawmakers passed the Standard Time Act, which sanctioned the four-zone system and authorized the Interstate Commerce Commission to define and make any necessary changes in the zones.

That same year, perhaps spurred by its ability to control a vague entity that so profoundly affected every aspect of life, Congress approved another time-related change: daylight saving time. A clock rendition of the calendar manipulations of Julius Caesar and Pope Gregory, it actually originated with Benjamin Franklin, who suggested that clock time could be advanced ahead as a way to take fuller advantage of daylight hours and save money on expensive candles. The idea was formalized in 1907 by an Englishman, William Willett, who, the story goes, was walking by a house with

the shades drawn even though the sun was high. Apparently taken by what he considered a dismissal of one of nature's blessed and valuable resources, he published a pamphlet called "Waste of Daylight," which inspired much of Europe to try and save it.

Willett's plan was a bit more complicated than the current system: He suggested moving the clock ahead 80 minutes in 4 turns of 20 minutes each Sunday in April. The hour advance won out, however, and was adopted first in Germany and England in 1916 as a way to conserve fuel during World War I. In England clocks were put ahead an hour for what was known as "summer time"; during World War II, "double summer time" advanced clocks two hours ahead of Greenwich time.

Daylight saving time was passed into law in the U.S. during World War I and observed for seven months in 1918 and 1919. It proved even more unpopular than standard time had been in its early days, and the law was repealed in 1919. However, a few states and such *Continued on page 152*

TIME, A MATTER OF MONEY AND SAFETY: For railroads, a schedule variation of even a few minutes not only inconveniences passengers but also can cause disastrous collisions, like the exhibition wreck (above) staged in 1896 by the Missouri-Kansas-Texas Railroad. A pileup on a track (right) in turn exacerbates schedules, creating long delays up and down the line.

TRACKING TIME:

Railroad conductors and their pocket watches once helped keep the nation's time synchronized by carefully clocking train arrivals and departures. Beyond mandating that conductors carry precision pocket watches, the railroads also spearheaded the movement to standardize time by dividing the country into time zones (map). Adopted in 1883, the U.S. plan was soon followed by a sequence of 24 time zones worldwide.

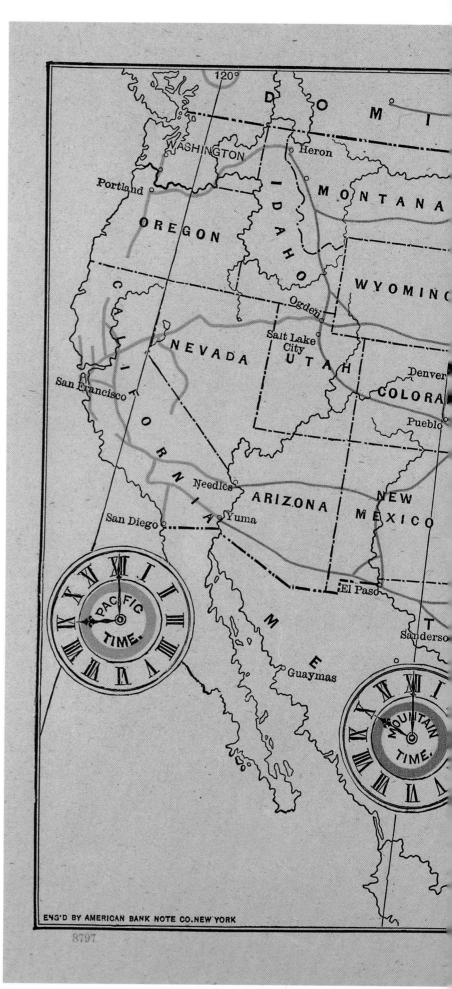

ENG'D BY AMERICAN BANK NOTE CO. NEW YORK

8797

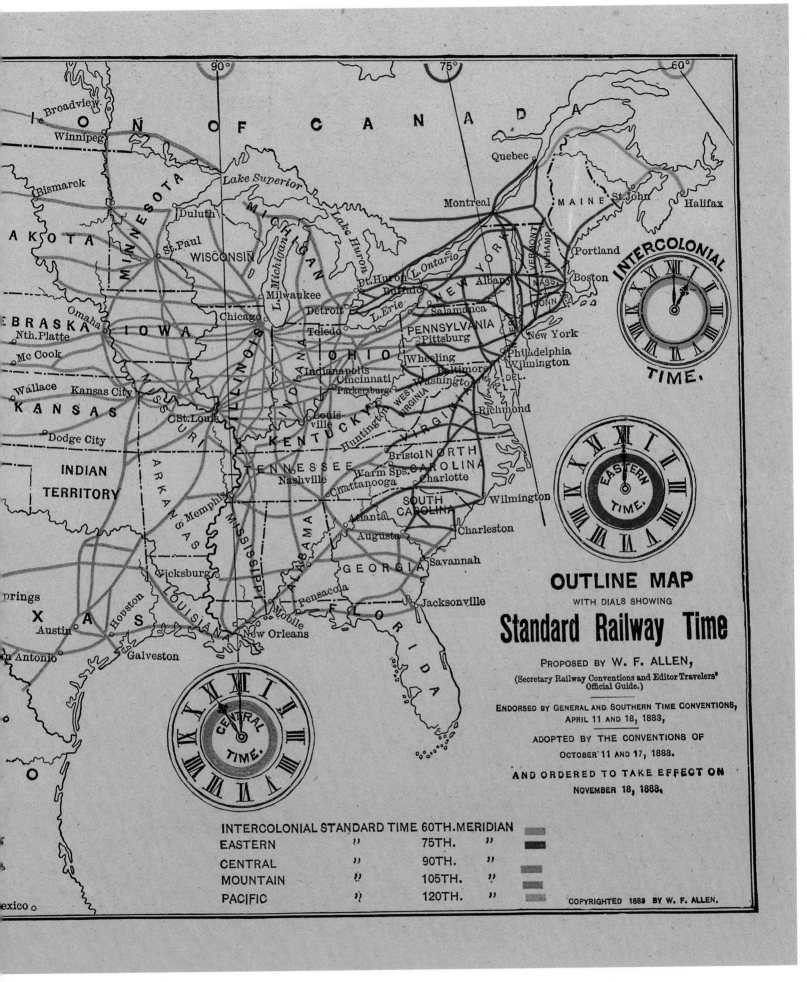

OUTLINE MAP

WITH DIALS SHOWING

Standard Railway Time

PROPOSED BY W. F. ALLEN,

(Secretary Railway Conventions and Editor Travelers' Official Guide.)

ENDORSED BY GENERAL AND SOUTHERN TIME CONVENTIONS, APRIL 11 AND 18, 1883,

ADOPTED BY THE CONVENTIONS OF OCTOBER 11 AND 17, 1883.

AND ORDERED TO TAKE EFFECT ON NOVEMBER 18, 1883.

INTERCOLONIAL STANDARD TIME 60TH. MERIDIAN		
EASTERN	"	75TH. "
CENTRAL	"	90TH. "
MOUNTAIN	"	105TH. "
PACIFIC	"	120TH. "

COPYRIGHTED 1883 BY W. F. ALLEN.

TIMELY TRADITION tops Britain's Royal Observatory in Greenwich (below). Installed in 1833, the observatory's time ball was for decades hoisted to the top of its mast every day at 12:58 p.m. and dropped at exactly 1 p.m., allowing navigators on the Thames to set their chronometers to Greenwich mean time. Just north of the observatory runs the imaginary line of the prime meridian, illuminated by a string of light bulbs stretching across Greenwich Park (opposite). In 1884 Greenwich became home of the prime meridian—0° longitude —the place from which time and longitude can be reckoned.

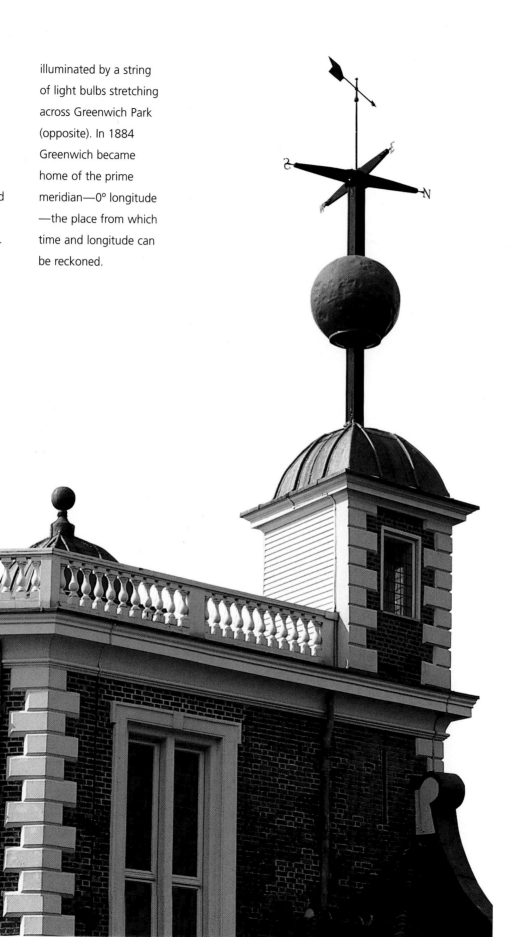

big cities as New York, Chicago, and Philadelphia, continued the practice.

Shortly after the U.S. entered World War II, the entire country was officially placed on daylight saving time and continued on it until the war's end in 1945. For the next 20 years, states and the communities within them were allowed to choose not only whether they wanted to observe daylight saving time but also when to begin and end it. The result was another burst of timely confusion. Some commuter railroads operated on daylight saving time and connected with other trains on standard time. Station clocks often displayed both times, but Connecticut once had a law that made it an offense to show any time other than eastern standard on a public clock. (Still, Connecticut's banks, offices, and shops observed daylight saving time.) Radio and television broadcasters continually revised their schedules to accommodate a regional timekeeping decision; train and bus timetables also needed regular revamping, and businesses that opened at, say, 8 a.m., on daylight saving time found they were unable to raise a similar business only a few towns away on standard time. Again, since time was money, something had to be done, and standardization to save daylight seemed to be in order.

Here, too, there were differences. Some businesses and industries, the railroads among them, supported daylight saving time, but farmers worried about what time the cows would come home, and drive-in movie operators about what time darkness would fall. Transportation interests organized a Committee for Time Uniformity, which fortunately hit upon an example that made a strong case for a standard: It found that bus drivers and passengers riding on a 35-mile stretch between West Virginia and Ohio had to suffer no fewer than seven time changes.

Congress again acted, this time in 1966 with the Uniform Time Act. Signed into law by President Lyndon Johnson, the bill set the last Sunday in April as the annual beginning of daylight saving time, when the clocks would be set ahead an hour, and the last Sunday of October as the date when they would be set back to standard time. Daylight saving time would now be uniform within each time zone throughout the nation, but, though uniformity was mandated, exemptions to daylight saving time would be allowed, so long as states passed laws expressing their preferences. Some did. Hawaii, the eastern standard zone in Indiana, and most of Arizona (except for the Navajo Reservation, which is under federal jurisdiction) do not observe daylight saving time. Nor do Puerto Rico, the Virgin Islands, American Samoa, Guam, and a few other U.S. Pacific islands.

In 1986, under President Ronald Reagan, the law was amended, dictating that daylight saving time would begin at 2 a.m. on the first Sunday in April, and, as before, shift back to standard time at 2 a.m. on the last Sunday in October. Energy conservation, again, was the reason for the spring change. An estimated 300,000 barrels of oil are saved each year by adding the entire month of April to the "daylight" months. (Some studies also suggest that crime, traffic accidents, and injuries are reduced by daylight saving time.)

SPRING FORWARD, FALL BACK is the mnemonic device that now keeps most of America alert to the changes that force time to loosen its possessive hold on the sun, allowing us to appreciate the light a bit longer. Daylight saving time, now observed in some 70 countries, is, of course, an illusion, in that it does not actually alter time, only the clock time for sunrise and sunset. "Losing" an hour and "gaining" one are but figures of speech, as imaginary as time's presence is on a clock. Since it creates a

WAKE-UP CALL:
Backed by the power of time, the unassuming household alarm clock can stir sleepers, drive them from their beds, and send them on their appointed rounds. These days, the familiar, teeth-rattling jingle of the older, geared alarm clock has been replaced in many homes by the gentle music of electrically powered clock radios.

uniform adjustment in the standard time of each time zone, everything still works out to 24 hours, even in time-distorted Indiana. There, 77 counties in the eastern standard time zone don't change to daylight saving time at all; ten others in the central time zone use both central and daylight saving; another five in the eastern zone rely on both eastern standard and eastern daylight saving.

THE CREATION OF TIME ZONES and daylight saving time contributed a lot toward a common understanding of time and its multiple uses. But even more important was the advancement in ever more accurate clocks and watches and their widespread availability. The key to their usefulness, however, is synchronization. It is essentially agreement on a point in time that exists in different places at precisely the same moment, or, put another way, a recurrence at the same successive instants of time. Synchronization occurs all around us: At airports, where air traffic controllers coordinate flight movements; in film studios, where technicians sync sound effects and dialogue with a film's action; in our television sets, where electronic beams assemble colored dots to produce an optical image; and in our own bodies, where the synchronization of internal organs keeps us alive.

By its very nature, because it has to do with things happening at the same time, synchronization has to be at the very heart of precise timekeeping. It is the essence of accuracy, the quality that modern clock-oriented societies demand. Without it, we would have to be satisfied, as were our ancestors, with telling time only to the nearest quarter hour, if that. It is what gives us the right answer—insofar as is humanly possible—to the familiar question, "What time is it?"

We take for granted that clocks need something to keep them in step with one another, no matter where they are. That something, as we have said, used to be a factory whistle, a time ball, or a cannon fired off at noon. Or it was a clock in a jeweler's window, relying on some form of local sun time, that set the standard—but only for a community and perhaps only for that jeweler's customers, who weren't aware that the jeweler across the street kept his own "right" time.

But before time is synchronized for use, it should be accurate. And timekeeping based on the motion of the Earth is, we now know, inaccurate, because the Earth's movement fluctuates in rate by a few thousandths of a second every day. That's good enough for most of us but not for scientists measuring everything from the speed of light through the vastness of space to the rush of electrons through microelectronic devices.

Today, time is determined, as we'll see later, by atomic clocks that measure the oscillation frequency of atoms and are accurate to an extraordinary billionth of a second a day. That "fine" time is disseminated through the satellites of the global positioning system (GPS) and serves as the ultimate standard for setting every wristwatch, every wall clock, computer clock, and radio telescope in the world. It's a far cry technologically from the day back in 1845 when the Naval Observatory in Washington, D.C.—which today maintains the Master Clock of the U.S.—dropped the first time ball in the nation. Set to the time extracted from careful solar observations, the ball was dropped from the telescope dome at the observatory's Foggy Bottom site every day at noon, the signal to synchronize clocks, chronometers, and watches in the city of Washington and aboard ships on the Potomac River.

Years later, when the time signals went national, they were transmitted by telegraph to communities with their own time balls. It was no scattershot

OUTRACING TIME, Western Union's female messengers roller-skated across waxed floors (opposite), delivering rush messages from various parts of the office to the main distributing center and saving up to nine minutes by their gliding. Outdistancing time, or just trying to keep up with it, has long been a human preoccupation. Overland Pony Express riders (left, upper) served the purpose from 1860 to 1861. They delivered the mail between Missouri and California in eight to ten days, until they were replaced with telegraph poles and lines. Far from having to hoof it across rugged terrain, Western Union's telegraphers, like the one in an 1873 engraving (left), transmitted Morse code messages across the country for prompt delivery to homes and offices as telegrams.

transmission: In 1877, the time ball atop New York City's Western Union Building was dropped on signal from the observatory, but even though there were no time zones, the New York signal came 12 minutes before that for Washington, to compensate for the longitude difference arising from New York's more eastern location.

Once again, the railroads played a key role in delivering time, as well as other goods. The Navy's time signals not only went by telegraph to cities and villages but via Western Union—whose main office clocks were electrically synchronized with the Naval Observatory—to railroads across the nation. Since the transmission of official time over telegraph lines was almost instantaneous, electrical or manual synchronizing of clocks and watches was now possible anywhere the wires reached. It was an involved process, one perhaps better explained by a former telegrapher, who practiced what is now the almost obsolete art of Morse telegraphy. Ed Trump of Fairbanks, Alaska, is one of that group, and his observations are indicative of the lengths taken to keep time in sync:

"These Western Union clocks were large, weight-driven pendulum types that were synchronized daily with the USNO signal, and tied into a 'time service' switchboard. This switchboard, or 'clock board' as it was known, was basically a switchboard terminating loop circuits with special relays, which were cut into Morse wires and city clock circuits to jewelry stores and banks. Contact closures in the master synchronized clock operated the switchboard clock loops, which in turn sent the signals out over the telegraph wires and clock circuits to the subscribers. At high noon, EST, the standard time signal was thereby transmitted to every office cut in on every telegraph wire....

"On the Morse wires, the time signal began a couple of minutes before the hour, and came across the wires as one click each second, until a few seconds before the exact hour, then a pause, and then a final click exactly at the The Hour. The wire would then close, and be available for regular work. Moreover, all railroad operations people—engineers, firemen, brakemen, conductors, foremen, dispatchers, and telegraphers—were required to carry a standard, railroad-approved, 21-jewel pocket watch (the only wristwatch allowed up until about 1975 was the Bulova Accutron), and an up-to-date watch inspection card at all times while on duty. You could count on being fired or at least suspended if you were caught working without these two items and a company timetable in your possession."

Western Union's sprawling clock circuit networks and the telegraph wires that supported them were in service until the 1970s, when there were still more than a few Morse operators and wires around. By the early 1990s, with the advent of new technologies, Morse code as a means of communication was virtually obsolete. The last function of the many Morse wires still paralleling the U.S. railroads was to transmit the time signals to all the rail offices. The thousands of watches—not to mention the Morse telegraph keys—that synchronized the arrivals and departures of thousands more trains are now collectors items, with mail order catalogues offering reproductions of them.

EVEN AFTER WESTERN UNION and the classic railroad watch had faded into memory, the need for more accurate, synchronized time had not. Long gone were the days when people had but modest requests of time. Clocks were now carefully set, not just hung and wound with their hands reflecting an hour, give or take a few minutes here or there. From 1904, the Navy was broadcasting its time signals by radio, basing them on ever more precise observations of the sun, stars, and planets. In 1910

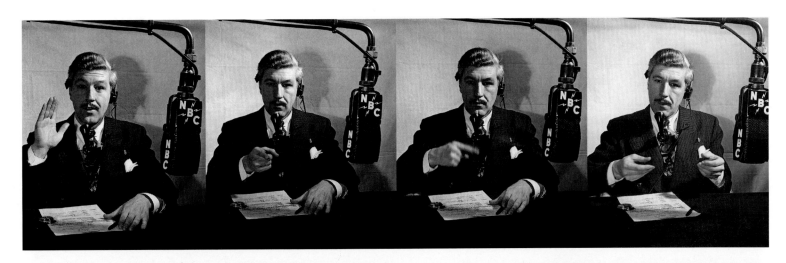

the French transmitted time signals from the Eiffel Tower and later came up with a "telephone robot" that automatically gave the correct time to callers who phoned a designated number.

By 1923, the National Bureau of Standards began providing round-the-clock shortwave broadcasts of time and frequency signals over its radio station, WWV. Much of everyday life had become seriously dependent on time—time that was foolproof. Scientists were especially needy. As they unraveled more and more of nature's mechanisms, their quest for more accurate and revealing measurements required that they rely on something better than the internationally accepted definition of the second as 1/86,400 of the mean solar day. If time were not used more efficiently—even its briefest intervals—industries involved in transportation, communication, scientific research, manufacturing, electric power, and a host of other pursuits would not be able to realize their enormous potential. As the 16th-century philosopher Jean-Jacques Rousseau, himself the son of a watchmaker, put it, "All that time is lost which might be better employed."

By the early 1900s, household electricity, with its steady flow of charge and signal frequencies measurable in cycles per second, was the logical next step in the quest for precision timekeeping. The

**ROUND-THE-CLOCK
WAR EFFORT**
keeps female workers
busy building nose
cones for World War II
bombers at a Douglas
Aircraft factory in
California (right). Intent
on making their own
patriotic contribution to
the war, women eagerly
joined the workforce,
replacing men who had
left to take up arms.
Though the pace could
be tough and time
moved fast on the
factory line, the women
kept smiling, particularly
when the noontime
factory whistle signaled
a welcome lunch
break (left).

keyword for scientists was frequency, a measure of events per unit of time. Frequency finds a place in just about everything old, new, and currently planned that keeps time. As one manufacturer of precision timekeeping and frequency control and calibration equipment contends, "Frequency can be measured just as precisely as time, because frequency is merely the inverse of time."

By the turn of the 20th century, clockmakers in Britain, Switzerland, France, and Germany had found several ways to use electricity from a battery to power their timekeeping mechanisms. But the electric clock we know today was an American invention, the creation of one Henry E. Warren, who saw the timekeeping potential of alternating current (AC) and used it in the patent of his "synchronous clock" in 1918. Unlike direct current (DC), which flows steadily in one direction, alternating current reverses its direction at regularly recurring intervals, moving its charges to and fro many times a second. Warren convinced producers of electric power to time alternating current cycles precisely, so that a synchronous motor could be used for a clock. It consisted simply of a small AC motor that drove the clock hands through a train of gears and told the time by the current—the 60-second alternating current in homes, controlled by the power plant's generator to a definite rate of alternations per second.

NOT LONG AFTER THE ELECTRIC clock appeared, inventors found an even more accurate timekeeper in, ironically, a common mineral that was the major constituent of the sand in the early clocks—quartz. The secret lay not in sifting quartz crystals through an hourglass but in what is known as a piezoelectric effect, a phenomenon discovered in 1880 by the Curie brothers, Pierre and Jacques. Essentially, it refers to vibrations produced when pressure is applied to a crystalline substance such as quartz, considered an ideal resonator material. Induced by a battery or alternating current, the vibrations then give off predictable, precise pulses of current. These, in turn, control the motor that turns a clock's hands or can advance the numerals displayed by liquid crystals in a modern digital display, a system of light-emitting optical devices that would have awed early gear, hands, and pendulum clockmakers by the absence of moving parts.

The first clock driven by a quartz-crystal oscillator was developed as a scientific time-measuring device in 1929 by researchers at the Bell Telephone Laboratories. Eventually, quartz crystals powered by miniature batteries found their way into watches, where their vibrations, at 32,768 times a second, were counted by an integrated circuit and then used to drive the motor that turned the watch hands or changed the digits on a display watch. Today, a finely crafted, expensive mechanical watch, accurate to within seconds every year, has found its match in an equally accurate, inexpensive quartz-driven watch.

Crystalline quartz's unique physical properties, including hardness and stability, also have made it essential to a range of modern electronics requiring precision timing and internal-clocking frequencies: Navigational equipment, radar, television, VCRs, microprocessors, and computers all use quartz crystals. Spare moments may be the gold dust of time, as someone once said, but the lowly crystals of quartz and the sophisticated mechanisms they drive actually count those moments. "At first glance," Bruce R. Long of the Piezo Crystal Company, has observed, "these useful devices appear simple, consisting of a slab of quartz between two electrodes, but their design requires knowledge of almost every area of classical physics, including mechanics, acoustics, wave motion, piezoelectricity, and

TINY TIMEKEEPER, a quartz crystal (above) metes out remarkably accurate time. A common mineral and the chief component in hourglass sand, lowly quartz revolutionized timekeeping. Its secret lies in its so-called piezoelectric qualities: It takes vibrations induced under pressure by an alternating current or battery and converts them into pulses of current. These unvarying signals, beating 32,768 times a second, are harnessed to turn a clock's hands or to advance the liquid crystal numerals in a modern digital watch (opposite). Even a cheap quartz-driven watch can electronic circuit theory. With careful design and manufacture, crystal resonators offer precision difficult to describe in everyday language."

But there was more to come. "What time is it?" was no longer always appropriate, certainly not for scientists. "Exactly what time is it?" was. But more to the point, so too was, "Exactly how long does it take?" Quartz clocks were doing the job better than most mechanical clocks, just as the pendulums and springs had outshone weights, but the quest for numerical precision had become a necessary obsession. As nuclear physics especially became more complex, the demand for more precise accuracy increased. Nuclear forces and models demanded accurate time measurements, particularly to monitor such illusive phenomena as the duration of so-called "magnetic moments" in an atom's nucleus. A Scottish zoologist, Sir D'Arcy Wentworth Thompson, probably characterized it best, saying, "The perfection of mathematical beauty is such that whatsoever is most beautiful and regular is also found to be most useful and excellent."

That regularity and usefulness was found, not surprisingly, in the atom, whose electrons oscillate as rhythmically as clockwork and can be made to serve as an invisible pendulum. The matrix of the material world, the atom with its spinning particles, was thus superbly qualified to connect with the immaterial world of time, providing a new form of clock that would lead to the splitting and redefinition of the second. It would be a tour de force, akin to splitting the atom itself—fragmenting the impalpable second into countable bits so tiny that time would no longer be doled out by huge markers running around a clock's face.

Of course, we know that instants are not made up of atoms of time, physical particles that can be fed into an atom smasher and pulverized, the fragments subdivided endlessly. Zeno of Elea, a fifth

century B.C. Greek philosopher, was the first to wonder how a specific interval could be composed of smaller intervals that could be made even smaller and smaller. Where would such intervals end? Some indefinable element with zero duration that exists—but for no time at all? It's hard to make sense out of such a paradox, but insofar as time researchers are concerned, the second lies somewhere in the realm of the infinite, able to be reduced to as many bits as technology permits. As one time theorist has said, "There appears to be no limit. Timekeeping is our own invention."

QUANTUM MECHANICS, the observation of ultra-small atomic particles, began the breakdown (metaphorically speaking) of time's traditionally smallest components. Austrian-American physicist Isidor Isaac Rabi led the way, inventing a technique known as magnetic resonance. With it, scientists can measure the natural resonant frequency of atoms. His discovery of new ways to measure the magnetic properties of atoms and molecules led to the development of the maser, an acronym for microwave amplification by stimulated emission of radiation. This device, the precursor of the laser, is used as a microwave oscillator or amplifier; it forms the basis of today's extremely accurate atomic clocks, precise to one billionth of a second a day. In contrast, digital watches with quartz-crystal oscillators are accurate only to one-thousandth of a second a day.

An atomic clock doesn't "tell time" in the usual way. Instead, it serves as a frequency reference standard for other clocks. Nor does it have a face in the conventional-clock sense. It resembles instead an elaborate piece of laboratory equipment containing two sets of electromagnets that separate atoms into beams, a mass spectrometer that selects atoms and directs them to an electron multiplier, and a quartz

match the accuracy of the priciest mechanical watches, themselves accurate to within a few seconds a year. Quartz crystal's other physical properties, particularly hardness and stability, also make it a reliable component in the electronic systems of navigational instruments, radar, televisions, VCRs, and microprocessors —all of which require precision timing.

crystal oscillator. The first one to be put into service used cesium atoms and was built at the British National Physical Laboratory in 1955 and required a roomful of equipment. Meanwhile, MIT physicist Jerrold Zacharias was developing an atomic clock small enough to be wheeled from one room to the next. Commercially produced, 50 Atomichrons were sold within four years.

The atomic clock's ticktock comes from measurements of the extremely regular waves of electromagnetic radiation. Emitted by the vibrations of atoms, they can be counted, somewhat as a pendulum's swing is counted. The resonant frequency of the atom thus is the consistent "beat"—like that of a metronome—that the atomic clock counts to record the time. It is all part of a natural system of interrelated rhythms in which time lies hidden.

"We are lucky to live in a universe having a large number of different processes that bear consistent time relations or frequency of occurrence relations to each other," Bradley Dowden of California State University at Sacramento writes in the Internet Encyclopedia of Philosophy. "For example, the frequency of a fixed-length pendulum is a constant multiple of the half-life of a specific radioactive uranium isotope; the relationship doesn't change as time goes by (at least not much and not for a long time). The existence of these sorts of relationships make our system of physical laws much simpler than it otherwise would be."

A variety of atoms—mercury, hydrogen, rubidium, and cesium among them—have the right moves to tell time by, and because their vibrations remain constant, these atomic markers measure

GOVERNED BY THE CLOCK, the pace of modern life makes even eating a fast-motion proposition. Now ubiquitous, McDonald's (above) pioneered the fast-foods concept in the 1950s with 15-cent hamburgers. Another pioneer of fast action, Federal Express is able to bring businesses "the world on time" by keeping its workers hopping (opposite).

short intervals of time much more precisely than a mechanical clock. And short is putting it mildly. While a second is still one-sixtieth of a minute insofar as most of us are concerned, that's not short enough for sticklers of precision like satellite trackers, navigators, and scientific researchers who need to measure microphysical events that occur over incredibly short spans of time. We're all familiar with stopwatches that can record time to the nearest fifth, tenth, or hundredth of a second, and with cameras that can capture fast time to a millisecond—a thousandth of a second—fast enough to create a stop-motion effect. But there is also the measurable nanosecond, a billionth of a second, two or four of which are what it takes for a personal computer to execute a single software command. The picosecond, one trillionth of a second, is one of the shortest periods of time that science can measure accurately.

The latest generation of atomic clocks—located in 30 countries and aboard each of the 24 satellites of the global positioning system—can keep time with staggering accuracy, neither losing nor gaining more than a second in three million years. Today, that second is defined as a frequency of resonance, that is, exactly 9,192,631,770 of the atomic oscillations associated with the cesium atom. The frequency of an electrical signal is measured in cycles per second, or Hz, for hertz. Millions of cycles per seconds are megahertz, and in the case of the cesium atom, its frequency of resonance works out to slightly more than 9,192 MHz.

But even the new definition of the second is not enough to keep time in sync. The problem is trying to link the *Continued on page 174*

RUSH-HOUR RUSH:
Whether they're moved by the spirit or by a job, people race against time to get where they have to be...on time. Their pace may be linked as much to where they are as to who they are. In a study of the hustle-bustle of American cities, psychologist Robert Levine of California State University at Fresno, found that Boston (left) topped the list, outrushing runners-up Buffalo and New York City. New Yorkers did lead in one category—the number of passersby wearing watches. But Madrid makes a commute in U.S. cities look like a stroll down Easy Street. Here, Madrilenos cram a bus for the final fight home. The city suffers through four rush hours daily, as workers toggle between home and work twice during the day: at midday for a siesta and again at close of business.

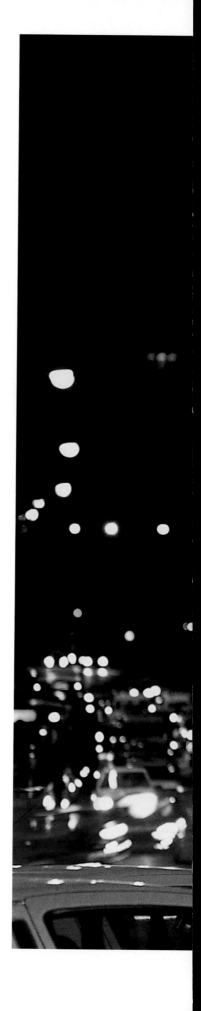

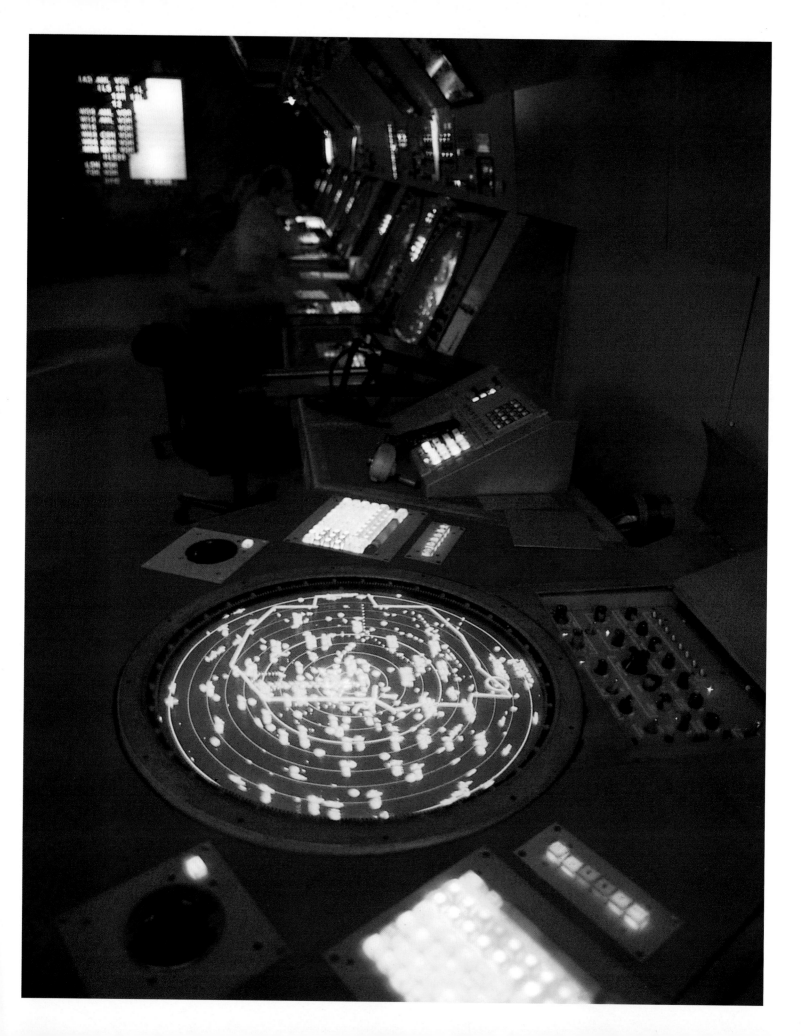

GLOWING WITH INFORMATION,
a radar screen has been
the heartbeat of the
Air Traffic Control Room
(opposite) at Dulles
International Airport,
outside Washington,
D.C., since the 1980s.
Invaluable for a
multitude of tasks, from
guiding aircraft to
catching speeding
drivers, radar detects
long-range objects and
calculates their positions
by measuring the time
it takes for radio waves
to travel to the objects,
reflect off them, and
return. Another
electronic device that
makes use of periodic,
electrical signals, the
oscilloscope (right)
works on the principle
that virtually all physical
effects can be converted
into electrical signals
and charted over time.
Its cathode-ray tube
produces a line graph of
those electrical signals,
tracking circuit variations,
seismic vibrations, and
other changes.

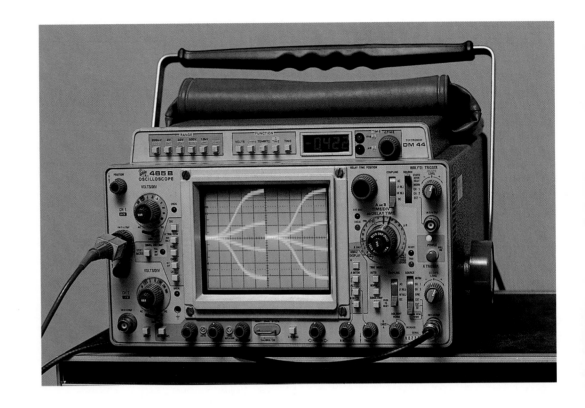

FROZEN IN TIME:
Piercing an apple at
1,900 miles an hour,
a bullet is captured by a
camera fast enough to
stop high-speed motion.
American electrical
engineer Harold E.
Edgerton pioneered
such photography, using
stroboscopic light to
study machinery in
action. The stroboscope
he developed produces
brief, regular flashes of
intense light; when the
frequency of the flash is
correlated to the speed
of a moving object—
a one-microsecond flash
can freeze the motion
of a speeding bullet—
an optical illusion of
slowed or stopped
motion is produced.
Stroboscopes are also
used to check moving
parts in machinery and
make adjustments
where necessary.

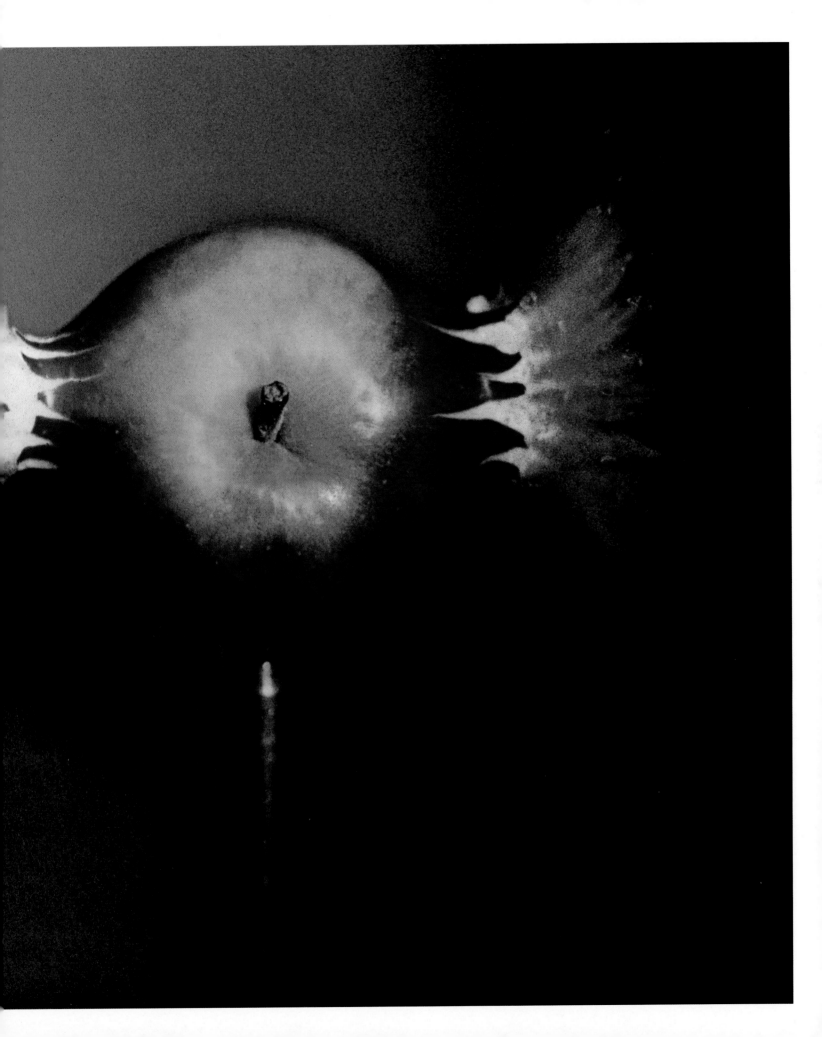

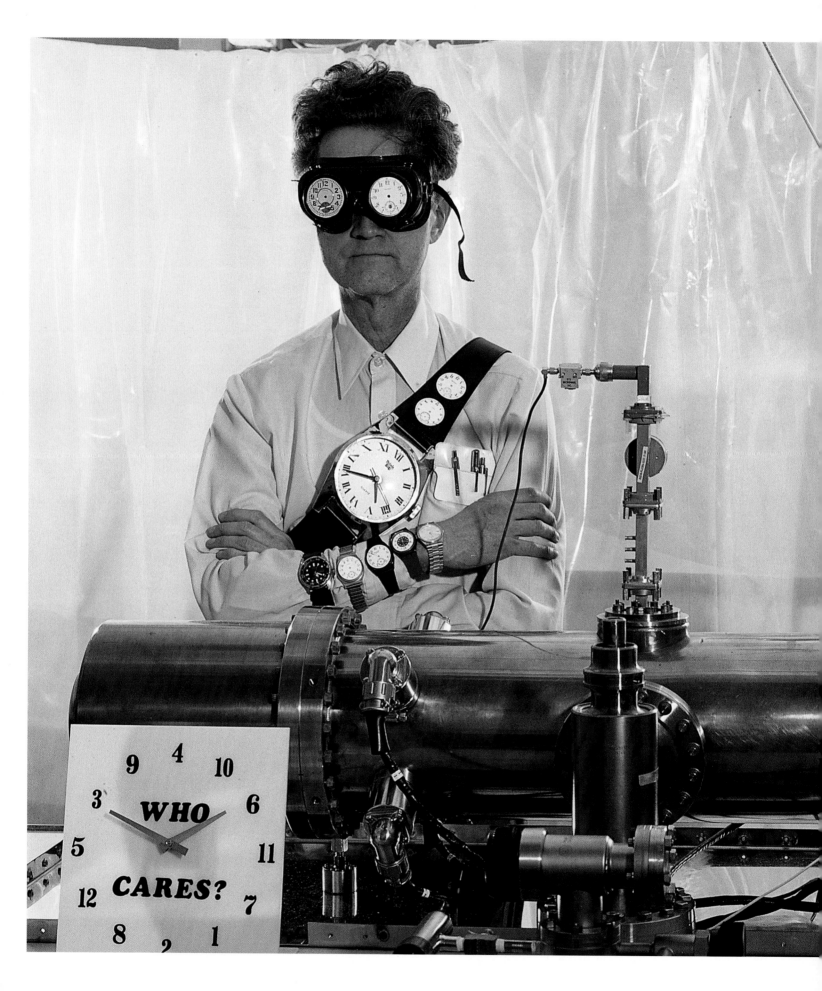

WHAT TIME IS IT?

Longtime National Bureau of Standards technician Jim Gray (left), jokingly called "Father Time," poses with the now obsolete NBS-4, one of the first atomic clocks to use cesium. Lacking the face of a conventional clock—and indeed any familiar clockworks—an atomic clock counts the extremely regular waves of electromagnetic radiation emitted by vibrating atoms. Infinitely more precise than a mechanical clock, an atomic clock keeps time with an accuracy that neither loses nor gains more than a second in six million years. At the clock's heart, a cylinder contains the metallic element cesium (right). A whirling dervish of the periodic table, the element's 9,192,631,770 atomic oscillations per second now serve as the standard by which the second is defined.

redefinition to Greenwich mean time. Based on the Earth's irregular motion, mean time is not uniform when compared to the time measured by atomic clocks. Atomic clocks, as good as they are, get out of sync with the Earth. In 1972, an international agreement rectified matters by allowing atomic clocks to run independently of the Earth, thereby keeping two separate times coordinated in a compromise time scale, appropriately named coordinated universal time, or UTC.

To keep things in order—that is, to keep the difference between Earth time and atomic time within nine-tenths of a second as the two times fall out of sync—leap seconds were inserted into the atomic time scale, resulting in 61 seconds in the minute just before midnight UTC. Between 1972 and January 1999, 22 leap seconds were added at intervals, because the Earth's rotation tends to slow down. If the Earth were to speed up, there would be a "negative" leap second, one that was removed, an event that has not yet occurred and which appears unlikely in the foreseeable future.

A solar day now lasts two milliseconds longer that it did in 1900, which means that a leap second has to be added approximately every 500 days. As the Earth's rotation continues to slow down, the gap between the duration of an "Earth second" and an "atomic second" continues to widen. Unless astronomers, physicists, and meteorologists can find a solution, leap seconds will have to be added more and more often.

CURRENTLY, "OFFICIAL TIME" is kept in the U.S. by the National Institute of Standards and Technology (NIST) and the U.S. Naval Observatory, both of which feed into UTC, now the world's universal time. The Naval Observatory is the largest single contributor to the international time scale, computed in Paris at the International Bureau of Weights and Measures. At the heart of the U.S. system is the Navy's master clock, accurate to better than a billionth of a second a day and based on more than 50 independently operating cesium clocks and a dozen hydrogen maser clocks. Frequency data from the ensemble are used to guide the frequency of the lead reference maser, master clock two (MC2), against which all other measurements can be corrected. The atomic clocks are spread over a network of more than 20 environmentally controlled "clock vaults" to ensure their stability. Every 100 seconds, intercomparisons of all of the clocks result in a highly reliable time scale and a clock reference system that produces the electronic time signals that in turn serve as the master clock.

Beyond calibrating most of the world's clocks, the nearly flawless atomic clocks also set the standard for coordinating an enormous global system of navigation. Though never part of the original plan when the clocks were invented, this network now feeds on all the refined time that the time providers can supply. And so, along with a correct answer to "What time is it?" we now have the answer to "Where am I?" and also "Where are you?"—all because of precision timing. Atomic clock accuracy is behind the satellite-mediated global positioning system (GPS) by which anyone with a handheld GPS receiving instrument can determine his position on Earth within a radius of 10 meters (32 feet, 9 inches) or less, a level of accuracy that had been reserved for the military until early 2000. The technology is based on the time it takes for electromagnetic signals to travel from satellites to GPS decoders on planes, ships, cars, even in hikers' backpacks. As each satellite in the system beams digital radio signals of satellite positioning and the exact time, a GPS receiver gathers the information from the satellites. Once the

CESIUM FOUNTAIN CLOCK

TIMEKEEPER OF THE FUTURE

Bearing a high-tech resemblance to the familiar water fountain, the NIST-F1 in the Boulder, Colorado, laboratory of the National Institute of Standards and Technology represents the eighth generation of atomic clocks. Inside, lasers propel clusters containing millions of cesium atoms upward, like a jet of water. Pulled back by gravity, the atoms fall through a microwave-filled cavity. When the microwaves are tuned to the right frequency, the atoms will fluoresce in the laser beams. Then the microwave frequency can be counted to keep time. While the clock's practical applications are as much as a decade away, its accuracy is proved—to within a second in 20 million years.

FARMING BY SATELLITE,
Doug Hartford (left) plots a course across his Illinois spread with the help of GPS—the global positioning system. Its constellation of 24 Navstar satellites, each with atomic clocks aboard, moves in six orbital planes around the Earth (right, upper); ground-based receivers calculate a user's position by measuring the time it takes for signals to bounce back from the satellites. The detailed maps (right, lower) generated by the system also help farmers identify areas of low yield to make more cost-effective applications of fertilizers. Aside from farmers, anyone with a hand-held or vehicle-mounted GPS instrument can lock into the system and determine his position on the globe within a radius of a few yards.

receiver is in touch with four satellites, the system takes over and computes the receiver's position in an instant. It does this by comparing its own time with the time transmitted by the satellites, using the difference between the two times to calculate its distance from the satellites. Without atomic clocks on the satellites, the receiver would not know exactly where the satellites were, nor would the satellites be able to keep the amazingly accurate time essential for the system to work flawlessly.

Today, telecommunications companies use GPS to synchronize their land-based digital networks, comparing their reference clocks directly with yet another kind of time, GPS time. GPS time and the GPS satellites that carry it monitor earthquakes and sliding tectonic plates, locate vessels lost at sea, keep track of truck fleets, direct the instrument landings of aircraft, survey the Earth, track forest fires, and even guide a bulldozer's blades. Installed in automobiles, a GPS system not only tells drivers where they are but also gives them directions to where they're going. In Japan alone, according to the U.S.'s National Academy of Sciences, a half million cars are already equipped with a GPS-based navigation system. That the time signals the satellites emit are not even in the same league as the signals that used to travel by a telegraph's tick is an understatement: The travel time of today's signals have an accuracy of ten nanoseconds, or ten one-billionths of a second.

One would think that a worldwide set of atomic clocks potent enough to render once hallowed Greenwich mean time obsolete—and accurate enough to become the guiding soul of virtually everything that depends on sequences of events, from radar to television to military missiles—would be enough for the timekeepers. But not so. Time, once elusive and stingy with its wares, is being pushed even further by new technologies vowing to

measure it, synchronize it, and disseminate it ever more efficiently. A new language has been created to deal with all of the obfuscation of time measurement research: "Aging" to mean "the systematic change in frequency over time because of internal changes in the oscillator"; "on time," which now means "the state of any bit in a time code that is coincident with the Standard Time Reference"; and "proper time," to mean "the local time, as indicated by an ideal clock, in a relativistic sense." One hopeful note: In the NIST's *Glossary of Time and Frequency Terms,* a clock is still defined as "a device for maintaining and displaying time."

JARGON ASIDE, physicists, guided by Rabi's atomic resonance ideas, have developed some novel variations in the endless quest for hyperaccurate time-measurement devices. For one thing, they've scrapped magnets in favor of lasers to select energy levels of the atoms that will do the timekeeping. Already in use and under further development, cesium atomic "fountains" employ laser light to launch atoms upward through laboratory versions of the clock towers of yesteryear. The fountains spew the spinning particles through what scientists call a "moving molasses" on a timekeeping mission that promises to improve time accuracy tenfold.

Destined for space and satellite applications, such clocks should set a frequency standard never imagined even in the wildest dreams of such early time trackers as Su Sung, Huygens, and Harrison. Atomic timekeeping's impact on the nightstand alarm clock may not be something we notice as we wake to music or a buzz. But the instantaneous timing it will provide to everything from radio broadcasts to telephone services to computer modems that respond to the click of a mouse will forever alter the classic definition of an "instant" as a portion of time too short to be estimated.

PENDULUM CLOCK WITH ATTITUDE, the first working prototype of the Clock of the Long Now stands eight feet tall and weighs in at a half ton. Created by engineer/inventor Danny Hillis and constructed primarily of monel, an alloy of copper and nickel, it tells the exact time and is designed to run for 10,000 years. As two numbered, outside rings on the clock tick off the years and centuries, its rotating face charts star paths. A mechanical computer and torsional pendulum, consisting of three rotating 22-pound tungsten balls, control the whole works, even recalculating dial positions and sounding an hourly chime.

CHAPTER 5

THE TIME OF
OUR LIVES

> "Forty is the old age of youth, fifty is
> the youth of old age."
>
> — FRENCH PROVERB

DOES A WATCHED POT BOIL? We know it does —eventually. But eventually is not precise, nor does it satisfy everyone's perception of how long it takes for a given boiling to begin. For some of us, the tea kettle takes forever to boil, if we stare at it. But when we leave it for a time and forget about it, it seems to summon us back all too quickly with a piercing whistle. For scientists, the water in the tea kettle will always reach the boiling point on a precisely timed schedule. Watching or not watching the process won't alter anything, no more than wishing time would pass can force a clock to slow or speed up its ticking.

These seemingly contradictory views of time as it plays out in the activities inside a simple tea kettle only reinforce time's chameleon-like nature, reminding us once again that there are many time scales and that one person's time is not always that of another. Time, according to physicist Albert Einstein, depends on how fast one is traveling and on one's frame of reference. We know that the time buried in whirling atomic processes does not mirror that measured by moving celestial objects. The time that depends on the way our minds perceive it and on the circumstances we find ourselves in is something like intuition: It is an indirect or ulterior view, one far removed from the visible and orderly time calibrated and synchronized by cesium clocks. Unlike the physical clock time that is kept by mechanical and electronic devices, personal time seems to need no control system other than the "eye of the mind," that same aperture through which our insight and intuition pass. Accessible only so long as we live, the private time conjured up by the mind's eye does not, once again, always appear the same to everyone. Like beauty, it is in the eye of the beholder.

When we think about the subjective perception of time, something is going on that it is seemingly at odds with what we know as objective clock time, the time that exists outside our consciousnesses and has been carefully measured out for us by calendars, trickling sand and water, swinging pendulums, uncoiling springs, vibrating quartz crystals, and the rhythmic oscillations of an atom's electrons.

It's tempting to speculate whether subjective time is somehow measurably real, some objective phenomenon. Does it vary in predictable ways from individual to individual, from circumstance to circumstance? Can it be charted? Or is it purely a figment of our imaginations, an illusion of variability, a sensation that is still governed, no matter how it appears, by the regularity of the sun, by the biological clocks that coordinate our circadian rhythms, and by all those other clocks and calendars that manage to measure the vague something we call time? The questions are bothersome, as bothersome as asking whose time is the right time, or whether time operates the same everywhere in the universe, or perhaps whether time could have existed at all before the universe was created.

Conundrums aside, the fact is, since we don't know all there is to know about time, we can't always explain why we sometimes experience time in different ways, nor whether our sense of it reflects any "real" movement or change.

As we've seen, time is not an absolute flow, as Newton believed; it is but a measuring tool that needs an event to be. It doesn't "go" anywhere, despite the mental trick that makes you think it does. It is analogous to the road in Vermont. Asked by a tourist where it went, a local farmer responded, "It don't move at all, stays right where it is, but it'll get you there." Thus, time is, as sociologist Victoria Koehler-Jones of the University of Nevada sees it, but "a social construction that brings order to our lives." We can also say that how the mind grasps time—whether it is still, slowed, or flying by—and the direction in which it seems to be arranged, is independent of a physics of time. Rather, our perception is influenced, sometimes distorted, by circumstances, by our age, gender, culture, religion, environmental conditions, body temperature, the deprivations of light and sound, even a mental state that is altered by drugs or hypnosis.

Each of us, it seems, can have his or her own

time system, a self-existing system that sometimes appears to run independently of everyone else's, depending on how we perceive and measure change. In that sense, our varying perceptions of time are real, as out of sync as they may be with the time on an accurate, objective clock. We may well possess some kind of physical regulator in the brain, a chemical pacemaker, that slows time for us when we watch a pot on the stove or sit in a dentist's chair, and speeds it up when we busy ourselves with some engrossing task. If this is so, there might be another powerful internal rhythm kept by a biological clock, one that not only regulates heartbeat, blood circulation, and sleep-wake cycles but also controls the rate at which personal time seems to pass.

For years, psychologists and chronobiologists have been intrigued by our sense of time and our varying perceptions of it—by our ability, without

benefit of clocks or other sensory assists, to estimate the time between events. Sometimes we are on the mark, other times we over- or underestimate time. But it is clear that we are not only aware of the present moment, even in a completely darkened room, but also able to sense some vague relationship between "before" and "later"; we seem able to intuit how much time "has passed" even without a visible succession of events to go by.

Children in particular have been a prime research focus of chronobiologists, and that's not surprising. Newborns, focused as they are on a seemingly never-ending present, obviously cannot make any connections between their moment of existence and what we call a past and future. An infant who reaches for a sharp knife probably won't remember that it even existed after his parents remove it; out of sight out of mind. Nor can he be told, "Wait a few

minutes while the bottle warms up." Even in their very early years, children give time no thought whatsoever, since their perception of reality is different—far more basic than that of us adults, who are so attentive to time, to counting off the seconds, minutes, and hours in an almost obsessive fashion.

While all organisms are subject to external and internal changes and to successions of sensations from the very moment of their emergence into the world, the precise reckoning of time, as distinguished from an innate, biological sense of time, takes a long while to learn. In children, conceptions of reality and time form gradually, through stages of exploration and through interacting with successive events and their environment. They learn to use crying to trigger actions by others, a sign of awareness that a past crying outburst resulted in, say, a bottle of milk. In effect, "I did it before and I got what I wanted." The past becomes embedded in their consciousnesses. So, too, does the future, which they come to see as an extension of the present. They begin to anticipate consequences. Now if they are simply shown a sharp knife or a hot iron, they back off. When confronted with a bottle of milk, they stop crying.

It gets more expansive with the arrival of gestures and later words. Around six months of age, psychologists say, babies start to realize that things exist even when they can't be seen, a concept that the famed Swiss pioneer in childhood development, Jean Piaget, called object permanence. By the end of the first year, a child will hunt for hidden objects and even demonstrate surprise if things aren't where they're supposed to be. "Object permanence and an infant's searching for hidden items both indicate that an infant possesses a mental representation, a schema in Piaget's terminology, for missing objects," one psychology text explains. "This is an important milestone for the child since it reflects the ability to recall prior events."

Soon, the child heeds the cycle of dinner, bedtime, awakening, breakfast, and play. He may not be able to see very far into the future, to the time when he will be as old as his parents, but he begins to recall some of his past. Still, it will take some time more before he is able to make the kinds of connections that will enable him, for example, to assess the relationship or the order of birth between his own or someone else's age. Piaget tells of one study in which an observer, after watching the spontaneous reactions of his daughter, concluded that young children tend to confuse age with height, as if aging were tantamount to growing. Piaget's own studies confirm how children confuse the concept of time with size, as demonstrated in the following dialogue with a child, age four.

"Have you any brothers or sisters?"

"I have a big brother."
"Were you born before or after him?"
"Before."
"So who is older?"
"My brother, because he is bigger."
"When he was small, how many years older was he than you?"
"Two years."
"And now?"
"Four years."
"Can the difference change?"
"Yes, if I eat a lot of soup I shall grow bigger than him."
"How can one tell which one is older?"
"The one who is bigger."
"Who is older? Your father or your grandfather?"
"They're both the same."
"Why?"
"Because they are as big as each other."

Between seven and twelve, children refine their concepts of time, along with those of numbers and probability. They speak in the adverbs of time, mentioning what they did yesterday and what they want to do tomorrow, not just what is happening now. As they grow older, that development of a time sense changes and sharpens. While an awareness of time builds, like knowledge acquisition, in a series of continual, sequential stages from birth, one's sense of it and how that sense is used may not be the same in every individual. Consider age, which is one of humankind's great examples of relativity. There is a vast difference between the way a teenager and an elderly person view time. Generally speaking, the higher the activity level of an individual, the quicker a given time span seems to pass; boredom seems to slow it up. By this standard, age apparently makes little difference: Time passes as quickly for a busy elderly person as it does for a busy teenager. But, perhaps because of an awareness that one's days are numbered, the way an elderly person assesses a given stretch of time does differ from the view of a young person who sees the future as too distant to care about.

AGE BRINGS WITH IT EXPERIENCE—the sum total of a long series of events stretching back in time; that experience, in turn, is accompanied by a greater, more intense appreciation and sensitivity to time. For older people, then, time passes, subjectively speaking, more quickly, and a time span that a young person might assess as long may be perceived as short by the aged. It may indeed be a long, long way from May to December, as Maxwell Anderson wrote, but the days *do* grow short when you reach September.

More than 60 years ago, Pierre Lecomte du Noüy, the distinguished French biologist and philosopher, expanded on the idea that young and old, although living in the same space, actually inhabit separate universes, where the value of time is radically different. While working at the Pasteur Institute in Paris, he was struck by the relationship of inward time—what he called physiological time—to wound healing. His studies showed a constant connection between the time it took for a wound to heal and the physical, or "calendar," age of the patient. His findings suggested that the rate of tissue repair was five times slower at the age of sixty than at age ten. The time of the body, thus, was not the same as the time marked by the clock. It was an aspect of the individual and not of the solar system. Moreover, du Noüy suggested, the rate of wound healing made it possible to derive the real physiological age of the patient, a unit of time different from the standard obtained from the rotation of the Earth around its axis. "Consequently, there are two kinds of time," he wrote in his 1937 book, *Biological Time.* "One corresponds to the classical notion, the

sidereal, physical time, without beginning and without end, flowing in a continuous, uniform rigid fashion. The other, the duration of our organism, which begins and ends with us, and which does not affect identically in our youth and our old age the phenomena of which we are the seat. It is a time which remembers. It is no longer the impersonal, rigorous time measured by the rotation of the earth, the immutable and arithmetical time in which the universe evolves. Its flow seems to be submitted to periodic fluctuations. It rebounds with each germ: a living time."

The main philosophical consequence of du Noüy's work was to postulate that time does not have the same value for a child as it does for a grown-up—that a year is, indeed, much longer, physiologically and psychologically, for a child. "One year for a child of ten corresponds to two years for a man of twenty," he wrote later in *Human Destiny.* "When the child is younger, the discrepancy is still greater. The time elapsed between the third and seventh year probably represents a duration equivalent to fifteen or twenty years for a grown man." There may be a plus side to this stretching of time in the young. With more psychological—but not chronological—time at their disposal, children have the capacity, du Noüy went on to say, to absorb considerably more knowledge during their first years than adults could ever absorb in the same time. "It would be highly desirable for parents and educators to take this fact into consideration," he concluded.

GENDER, TOO, APPEARS to affect one's perception of time. Women, some studies have shown, tend to be less proficient than men in coming up with an exact estimation of time and seem to be more inclined than men to overestimate. According to some, a reason for this might be that men time most of their activities by the regular movements of the clock,

NEWBORN TIME:

To an infant, time exists
as a never ending now,
with no past or future.
But the older we get,
the more we become
attentive to time and its
mystery, its passing and
its potential. Holding a
baby, we are keenly
aware that we ourselves
were once newborns,
and that, as time flows
forward, it will carry
the child to the place
we are today.

while women are influenced by experiences and cycles—menstrual cycles, gestation periods, and menopause—which are longer and not as precise as clock time. Whether this is true or an example of male chauvinism seeking to explain female behavior in purely biological terms needs a further test of time. If indeed women overestimate time, then they would not, as the questionable proverbial wisdom has it, always keep men waiting, because they'd instinctively be inclined to budget more time than they needed to get to a meeting and thus would be there quite early!

YOUTH, AGE, AND PERHAPS GENDER are not the only factors that have an effect on our time sense. Other, more specific agents are also sometimes at work, and these may be influenced by metabolism or some still obscure bioelectrical rhythm in the brain that brakes or accelerates our sense of time. It's logical to assume that subjective time's orderly succession of sensations can be warped by physiological and biochemical changes in our bodies, induced by such things as environmental conditions, drugs, illness, and stress.

Consider hibernation, the trancelike state in which bears, badgers, and a host of other northern animals pass the winter, fat with food, body temperatures lowered, and oblivious to the outside world. When their temperatures rise and they wake up in the warmth of spring, all they know is that it was relatively warm when they retired and that it is still warm as they awaken. Everything that passed in between, the winter sunsets, the snow and frost, never happened for them. It was as though time had been compressed and several months lost to their memory.

Temporal rhythms in other animals and in plants are also affected by temperature. Experiments on human subjects have shown that, when tempera-

tures are changed, they, too, can miscalculate the time between events. A classic experiment conducted some years ago by Hudson Hoagland of the Worcester (Massachusetts) Foundation for Experimental Biology provided striking evidence of a physiological alteration in subjective time. Dr. Hoagland's wife had been ill with influenza and was running a high fever of around 104°F. She sent him

"MY BOOK AND HEART" must never part," the old *New England Primer* proclaimed. But we can sometimes part from time and escape our sense of its

about a half minute to count to 60; but when he repeated the experiment as his wife's temperature began to drop, her counting pace slowed. And when her temperature returned to normal, so, too, did her estimation of what constituted a second as measured by the clock. The experiment was an indication that heat may speed up some chemical reaction in the brain, which in turn has an impact on our determination of time, in effect making it pass more quickly than a clock allows.

But Mrs. Hoagland's case, significant as it was as an example of how psychological time can be exaggerated by some physiological change, pales in comparison to the time contractions induced by certain drugs. Perhaps the most graphic example of a drug's effects on time perception is found in the English author Thomas De Quincey's autobiographical narrative, *Confessions of an English Opium Eater.* First published in 1822, the work is told from an addict's point of view—De Quincey remained one until his death. Along with his accounts of the pleasures of a drug habit, there are detailed reports of a vastly accelerated inner time that occurred when he was under the influence of opium. Describing the shadowy dreamworld of his own subconscious, De Quincey said this: "The sense of space, and in the end the sense of time, were both powerfully affected. Buildings, landscapes, etc. were exhibited in proportions so vast as the bodily eye is not fitted to receive. Space swelled, and was amplified to an extent of unutterable and self-repeating infinity. This disturbed me very much less than the vast expansion of time. Sometimes I seemed to have lived for seventy or a hundred years in one night; nay, sometimes had feelings representative of a duration far beyond the limits of any human experience."

Opium is no longer the recreational drug of choice, but some of its psychoactive successors have the same power to warp time for the user, some-

perpetual passing by losing ourselves in a book—or in the comforting familiarity of daily routine—as this family is doing.

to pick up some medicine at the local drugstore, and, when he returned 20 minutes later, she chided him, insisting he had been gone for at least an hour.

The time discrepancy fascinated Hoagland, and he decided to experiment by asking his wife to count to 60 at a rate of what she believed to be one figure per second. When Hoagland timed his wife with a stopwatch, he found that she had taken only times making it longer, sometimes shorter. The hallucinogens marijuana, LSD, and mescaline produce states of altered consciousness and can make time race for some users, or stand still or run backward for others. Stimulants like caffeine, cocaine, and amphetamines do just what the slang term "speed" implies: They speed up some bodily activities through their effect on the central nervous system and make interval time seem longer, forcing the user to overestimate it. Depressants like alcohol, barbiturates, and tranquilizers slow things down and make time seem shorter, which causes the user to underestimate.

Hypnosis, too, which heightens suggestibility and enables one to ignore distracting events, can skew the perception of time. Its effects can be as dramatic as those produced by drugs or changes in body temperature, perhaps more so since it involves the transfer of one's conscious control to another person. In a hypnotic trance, the sense of time can be as non-existent as it is in a dream, or it may be manipulated by the therapist so that the hypnotized individual accepts a suggested time scale—one in which, say, the tick of a clock's second hand seems much slower than it is. A few minutes can become hours, events may be compressed, lost, or paraded in slow motion across the subject's mindscape.

Such bizarre effects do not always require a hypnotist. Time-altering hypnosis, most of us have found, can also be self-induced by relaxation techniques, by concentrating on one's breathing, by flashing lights, or by a variety of monotonous practices and recitations. Our minds wander during a boring lecture and our eyes glaze over. If we were asked how long we were out of touch with reality, most of us wouldn't know. Or we might sit rapt through a two-hour concert but not be able to estimate the duration of the individual pieces we heard. Likewise, after the trancelike feeling of highway

hypnosis experienced during a long, boring drive, we're unable to recall details of the trip, tell how long we've been on the road, or how long it took to get from one point to another. Amnesiac states produced by brain injury or disease, acute conflict or stress or hypnosis can also adversely affect one's time estimation. Typically, individuals with impaired memory or complete memory loss may grossly underestimate the time in which they have been engaged in some activity, or they may overestimate the time that has elapsed since a past event occurred.

The peculiar rhythms of subjective time are also affected by illness, both physical and mental. Intolerable pain, even if brief by clock time, seems to stretch the present to such an extent that a moment becomes meaningless to the sufferer. Even when it's over, time sense remains altered, since we often cannot precisely re-experience the pangs at all. Schizophrenics, who can live in a terrifying world of roller coastering moods, where people and familiar objects seem like reflections in a distorting mirror, are especially susceptible to twists of temporal judgment. Sometimes their minds seem to race so fast that they cannot brake them; on other occasions thoughts creep along at a maddeningly slow pace.

Some years ago, Curt B. Richter of the Johns Hopkins School of Medicine collected medical records of some 500 patients and found that during illnesses an inner clock mechanism ticked at its own pace and governed symptoms and the progress of the disease by its own, personal system of measurement. One woman with Parkinson's disease, for example, was unable to speak coherently or walk or feed herself, except at 9 p.m. every day. As the hour neared, her condition improved dramatically, and she was apparently normal for three hours. "Similarly," said one account of Dr. Richter's compilation, "a paretic who suffers from progressive dementia and paralysis may believe he *Continued on page 199*

Continued on page 199

CHRONOBIOLOGY
'TWEEN TIME, TEEN TIME

A study that was conducted jointly by the Yale Clinic of Child Development and the Gesell Institute of Child Development, both in New Haven, Connecticut, showed the following differences in time perception among children 10 to 16 years old.

Age 10 sees time as "something the clock tells," a definition related to specific units, like minutes, hours, and centuries. There is not yet much feeling for the passage of time, but there is a sense that there can be differences in how long the time units seem, as in "Good things go fast, bad things go slowly." "Ten's own management of time is so improved," the researchers commented, "that it is a great relief to everyone concerned, especially his parents."

Age 11 has a sense of the relentless passing of days, minutes, hours. As one 11-year-old put it, "No matter what you do, even break all the clocks, you can't stop it."

Age 12 is less confident that she can exactly state her ideas of time and instead defines it in terms of durations or spans: "A period of, well, time." "The length between something and something else." "Time is what you measure life in." Time drags very little for 12-year-olds, because they fill it so full and generally organize it into manageable blocks.

Age 13 "with his probing, defining intellect is less at a loss

to define time than he was at twelve." He now shifts from a fluid concept of time to a more static concept, seeing it as the "space between one event and another."

Age 14 most often defines time in terms of action and movement: "Things grow, things die, anything can happen." He is more oriented toward the near future and enjoys the present. He does, however, "look forward to big things in the future." But control of time at this age seems to vary greatly, and often the 14-year-old seems to be constantly rushing and often late.

Age 15 defines time as a space and an abstraction, but one that is still "something very tangible." As another put it, "You can't count on it, you can't figure it out, and yet it goes on endlessly like a stream." Though the 15-year-old is in better control of time than 14-year-olds, she would still "rather be slightly rushed than sitting around doing nothing."

Age 16 offers the largest personal variety in the definitions of time, and yet 16 still sees time in terms of action and processes: "The moving onward of things," "Life slipping away." Sixteen-year-olds also think of time as a measurement or way of recording, as in "the measurement

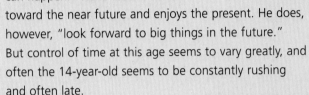

of the order of events," or "a way of recording events in a period." Their concepts also express inner states: "Days as they go by," and "Going forward." Moreover, they have now become aware of other people's sense of time and manipulate their own time sense to agree with that of others, so that, on the whole, they are usually on time.

AN IBO GIRL

in the Democratic
Republic of the Congo
flaunts the white body
painting that signals
the onset of
menstruation, a joyous
event in her culture
associated with one
of life's transitions.
Rites of passage mark
the individual's move-
ment through life,
recording the passage
of time and changes in
social status—from
puberty to maturity,
from single to married,
even from life to death.

INITIATION RITES,

from acts of purification and instruction even to acts of mutilation, usher youths into adulthood. At the 2,000-year-old Shwedagon Pagoda in Myanmar, a young boy, decked out in elaborate finery, contemplates the infinite compassion of Buddha as part of his *shin-pyu* rite of initiation. A key to the future, rituals celebrate the moments at which new generations move through the ever-repeating panoply of life. Preparation for marriage, endurance, courage, death, resurrection—all are themes of the many ceremonies of adolescence.

SACRED TIME: In a Good Friday procession in Peru, worshippers honor the death of Christ and anticipate his resurrection on Easter Sunday. For many Westerners, particularly those raised in the Judeo-Christian tradition, time moves forward linearly, a transition from birth to death, a passage that is future oriented. But for Buddhists and Hindus and other adherents of Eastern religions, time moves cyclically, on an eternally circular path where events recur again and again, and rituals cele-brate their predictable, perpetual repetition.

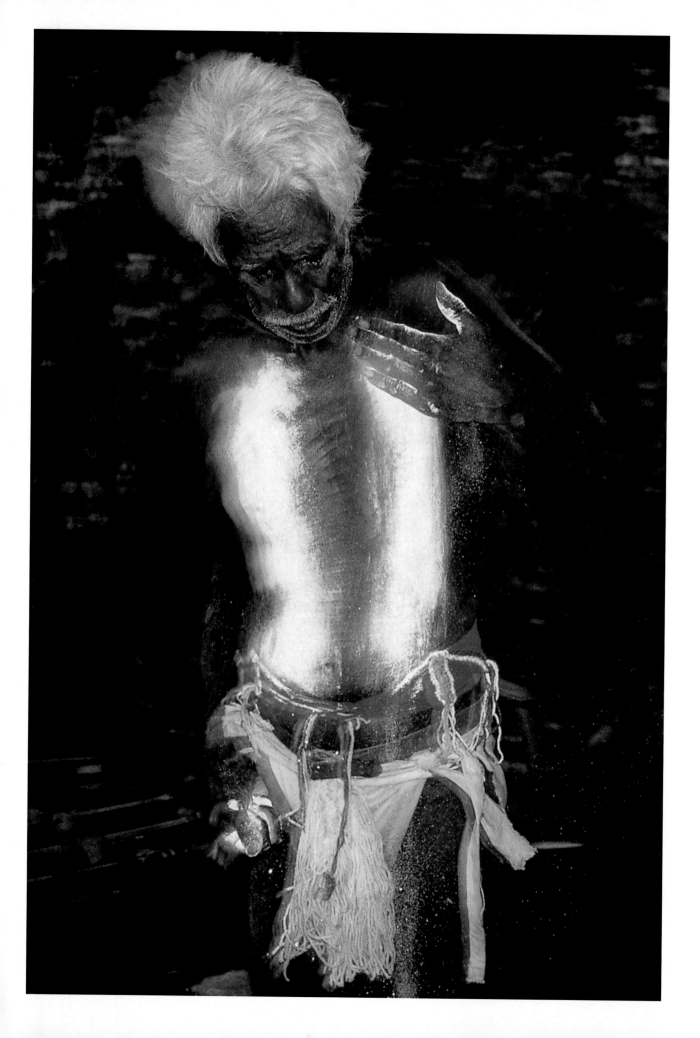

DREAMTIME,

the mythic past of Australian Aborigines, is celebrated by a tribal elder in a ritual called a corroboree (left). Smeared with white plant pigment, he recalls his own origins and the spiritual forces that granted energy to his life. Peopling dreamtime, the Aborigines' totemic ancestors— either spirit Sky Heroes or humans with supernormal powers— established the world and its customs. At right: In his own drug-induced dreamtime, a Hmong farmer in Asia's Golden Triangle, along the Mekong River, prepares to drift off with another bowl of opium. Like other drugs, as well as mental and physical illnesses, opium can distort one's sense of time, either accelerating or slowing it, even seemingly stopping it.

has remained the same age since he first became infected with syphilis. Time for him has stood still." In another case, a victim of Korsakoff's syndrome, which is a chronic delirium usually arising from alcoholism, "was likely to suffer impairment of the sense of 'pastness' of his experiences that is essential for normal recollection and the feeling of personal identity." Thus, for the ailing, as for the patient on drugs or under hypnosis, time becomes obedient to some inner clock; it becomes independent, flexible, and transformed into a phenomenon often markedly different from the time most of us live by every day.

But there is more to time disorder than inaccurate perceptions and estimations of measured time. Such things are but evidence that the rate at which time seems to pass is not uniform. Real alterations in time—and in the light that gives us a clue to its orderly presence—affect us more directly, through

sleep-wake disruptions in our biological clocks, in our circadian rhythms. The most common causes of real-time disorders are jet lag, changes in work schedules, and the condition known as seasonal affect disorder (SAD). These disorders result from a decrease in the measured, natural light to which we are exposed. In such instances, time is not just seen in a different way but becomes capable of muddling up our sleep, our moods, our productivity; on occasion it even sets us up for injury.

Consider jet lag, a disorder familiar to modern-day travelers who cross time zones. It is a perfect example of an intricate relationship gone awry, the one between sleep and time. Normally, time gives up some 8 hours of its 24-hour day so that mortals can take a break, soothe their souls and bodies, and prepare for work again. But, although jet lag noticeably disrupts sleep, it has many deeper ramifications,

such as effects on body temperature and water and electrolyte secretion. Travelers typically suffer from fatigue, gastrointestinal complaints, and shortened attention span—all the result of a conflict between the passenger's circadian clock with its temperature and other cycles in one time zone and environmental cues, like light, in another. Though the sleep-wake cycle eventually catches up to the new time, adjustment periods vary depending on individual makeup, the number of time zones crossed, opportunities for rest before and after the flight, and even the direction of the flight. (Travelers going west are not as likely to have trouble sleeping at night in the new zone but may experience morning insomnia; traveling several time zones to the east may mean evening insomnia and late awakening.)

Jet lag, fortunately, is a temporary mix-up. Travelers can cope by resting before and after a trip and by readjusting their eating schedules. But shift-work schedules, a mainstay of modern industry affecting 20 percent of American workers, are not easily dealt with. Unlike jet lag, the effects of shift work—jobs that have night shifts or require shift rotations or odd hours—are prolonged, sometimes for months and years. Humans are, after all, creatures who are generally active in the light of day. Altering that time-honored schedule throws the body and mind into a state of desynchronization, causing not only sleep disruption but fatigue, health problems, poorer job performance, a proneness to error, even domestic difficulty.

Mood may also be affected. Recently, researchers at Harvard Medical School found that changing natural sleeping patterns and hours of wakefulness produced changes in mood that varied with the time period subjects were awake. "When sleep is displaced, as in shift-workers," the researchers said, "the phase relationship between the sleep-wake cycle and the circadian pacemaker changes, and this change may alter mood during the waking episode."

Mood change is not the only adverse effect, however. Compromised performance—and, worse, serious and sometimes fatal accidents—often result because of fatigue and sleep deprivation due to shift work. Any variation of mental alertness can produce unwanted results in a multitude of occupations, whether the tasks require precision or not. Imagine what would happen if gymnasts, divers, and trapeze artists, who rely on precise timing for their "blind" maneuvers, had their inner clocks disabled because they suddenly had to work unaccustomed hours. Fortunately for these occupations —as well as for that of diamond-cutter, microsurgeon, airline pilot, and demolitions expert—shift work doesn't usually hold the same meaning it does for people in manufacturing, health care, and transportation, all industries that maintain operations around the clock.

Truck driving is a case in point. According to the National Transportation Safety Board, the probable cause of death most frequently cited in heavy truck accidents fatal to the driver is sudden or accumulated fatigue due to long working hours. Indeed, most accidents occur between 4 and 7:30 a.m., which coincides with the time that drivers are most likely to fall asleep at the wheel.

Sadly, among the worst casualties of muddled work rhythms are hospital patients inadvertently neglected because of overworked medical staffs—especially doctors in training, who put in excruciatingly long hours. In one celebrated case, a young woman died eight hours after she was admitted to an emergency room. A grand jury looking into the incident determined that the 18 straight hours worked by both the intern and resident before they treated the woman contributed to her death.

FOLLOWING PAGES
HEALING LIGHT:

Clients at a health and fitness center in Aachen, Germany, don high-tech versions of sleeping caps that beam ultraviolet light into their faces. Exposure to bright light seems to alleviate seasonal affective disorder (SAD), caused by deprivation of a time cue like sunlight. The disorder can leave sufferers sleep-deprived, depressed, and afflicted by a skewed sense of time.

Lost sleep, even during the seemingly innocuous switch to daylight saving time, can also have adverse effects. Though not as severe as the long-term results of shift work, that one-hour spring forward most of us experience every April can have, according to some researchers, the same fatiguing effect as a three-hour jet lag. A 1998 report in the *New England Journal of Medicine* by psychologist Stanley Coren noted a 7 percent increase in traffic accidents the day after DST goes into effect, and a 7 percent decrease in the fall when the clocks returned to standard time. "We're all sleep deprived anyway," Coren observed, "so that extra loss you experience is enough to lead to an accident."

Seasonal affective disorder is another compromiser of productivity and mood. Often incorrectly referred to as "winter blues" or "holiday blues," it affects sufferers in the autumn and winter seasons, when there is a decreased exposure to sunlight. It leaves its victims sad and depressed, lethargic, sleep deprived, irritable, and socially withdrawn. Some 11 million people are said to be afflicted with full-blown SAD (although a majority of the U.S. population experiences some mood change as seasons rotate), with women four times more likely to be affected than men.

Although the trigger mechanism for SAD appears to be light—indeed, exposure to bright light, especially in the morning, seems to improve the symptoms—scientists have recently homed in on a gene that may indicate susceptibility to the disorder. The gene apparently affects the action of the neurotransmitter serotonin. This important mood-regulating brain chemical is inhibited as less light passes through the eyes during fall and winter months. But, while there appears to be a genetic component, another mechanism may be at work, one that suggests sufferers are only responding to some other time cue of nature. According to Leo

Sher, a scientist at the National Institute of Mental Health, it may be that humans are programmed to sleep during the winter; we only force ourselves to stay awake by modern, artificial means. "These patients," Sher suggests, "may simply be biologically closer to people who lived in the 16th and 17th centuries, before we had electric light and central heating."

ANY DISCUSSION OF PERSONAL time isn't complete without mentioning culture and religion, forces that shape not only our thoughts and actions but also our sense of time. Beliefs and traditions are very much behind the different ways we view and use time and the direction we believe it moves in. Now, common sense seems to indicate that time, even though we know it to be as amorphous as space, reaches out, unlike space, in one direction—forward, into the future like an arrow. Our calendars and clocks convince us of this, and so, too, does the psychological sense that time passes and thus allows us to remember the past and not the future. (We'll talk about another arrow of time, the forward- and backward-running time of the subatomic world, in the next chapter.)

This linear view of time is imbedded in Western, Judeo-Christian thinking. In this view of the world, time is associated with sequential events; with human progress and the forward movement of history; with the transition from birth to death; and with the creation of a universe that has an unrepeatable beginning, an outward expansion, and a definite ending. Personal time, as many of us in the West see it, is a one-way track to something, often to something new and often without a glance backward. We go from elementary school to middle school to high school and through college. We take on a task and bring it to a conclusion. We play a sport, and in the end win or lose the game. As

HYPNOTIZING TIME:
Under the influence of hypnosis, which can be induced by concentrating on a device like the hypnodisc (left), a patient may have no conception of time at all or will accept the time scale suggested by the hypnotist. Time-altering hypnosis can also be self-induced by relaxation techniques, concentrating on one's breathing, performing monotonous tasks, or participating in group self-hypnosis. Using the latter technique, participants in a 1930s seance (opposite) lose all sense of time as they attempt to communicate with the spirit world.

sociologist Koehler-Jones sums up our concept of time, "Consider the dominant Western view: future oriented, active, and individualistic…. Popularly, we subscribe to a linear, monotonic progression of time from the past, through the present, into the future, but we have little use for the past. Our values are largely near-term while we concentrate on the immediate growth ahead. Our desire to continue economic abundance encourages us to overload the present and near future with planning."

Those of us who adhere to an almost ritualis-tic, linear time live in a system that anthropologist Edward Twitchell Hall has called monochronic time. Monochronic people, he observes, typically have short-term relationships, do one thing at a time and concentrate on the job at hand, stick religiously to plans, emphasize promptness, and seldom borrow or lend. But there is another time system, a polychronic one, that is not so rigid.

Polychronic people, many of whom live in Asia, Africa, the Middle East, and South America, may not pay as much *Continued on page 211*

"NAE MAN CAN TETHER TIME or tide," warned Scottish poet Robert Burns. And yet, when it comes to time, we often try desperately to alter its pace...and its effects on us. While teen-age girls apply make-up (right) to look older, older women apply mud packs (opposite, lower) to help them look younger. The elderly, for whom time runs too fast, may feel younger in body and mind than their years confirm. Still doing headstands at 79, a performer (opposite, upper) with the Sun City Pons, the dance team from an Arizona retirement community, stands time on its ear.

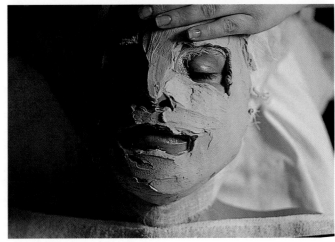

"LONG LIVE DEATH," reads the plaque below a diorama crafted for Mexico's Day of the Dead celebration. Held in early November on the Catholic Church's All Souls Day, the holiday makes light of death with smiling skeletal images and graveside feasts. Waiting patiently at the end of every life, death is celebrated with its own special calendar day in many cultures. Some observances honor the dead, others placate the ghosts of the departed to keep them from pestering the living.

attention to the clock. They do many things at once, change plans often and easily, base promptness on the relationship, tend to build lifelong relationships, borrow or lend often and easily, and consider deadlines objectives to be achieved, if possible. Such a "cyclical" view of time is an ancient one. Central to Buddhist, Hindu, and Taoist beliefs, it also held sway in the West through the early Greek and Roman periods. In it, time is on an eternal, circular swing: Things that once were recur again and again, like the seasons. The Roman emperor and Stoic philosopher, Marcus Aurelius, put it this way, "What follows is ever closely linked to what precedes; it is not a procession of isolated events, merely obeying the laws of sequence, but a rational continuity. Moreover, just as the things already in existence are all harmoniously coordinated, things in the act of coming into existence exhibit the same marvel of concatenation, rather

than simply the bare fact of succession." One current writer on Asian thought and culture expresses it in more modern terms: "In such a scheme of endless periodicity final goals are bound to be more abstract than…the Heavens and Hells of the Christians, Jews, and Muslims."

Chances are most of us view time both as linear and circular, depending on circumstances, and any attempt to characterize one society is an oversimplification. Stereotypes do abound, though: Germans and Japanese are obsessed with punctuality; Brazilians are polychronics, people who feel that missing an appointment or being late for one is acceptable behavior. But a recent study presented at a British Psychological Society conference in London found that, although there was some evidence in support of such sweeping generalizations, other evidence contradicted them. The study's principal researcher, Peter Collett of the University

REBIRTH OF A SOUL:
High in the Himalaya, past and present merge in the person of Tsering Dorje. The young Sherpa boy, born in 1988, is believed to be the reincarnation of a lama who died more than a decade ago. To Buddhists, time—and human souls—cycle over and over again: What is happening now and will happen in the future follow inevitably from what has been.

of Oxford, told of asking managers of European countries whether they were concerned about lateness, whether they expected an apology for it, and, finally, how late is late. His efforts revealed that, although behaviors in general are consistent with popular stereotypes, people are frequently less concerned about lateness in others than in themselves. Punctuality, it seems, comes down to one's own perception of time and to one's definition of and attitude toward being "on time."

According to Collett's research, which appeared in the British journal, *Nature,* German managers, for instance, are quite tolerant concerning tardiness. To them, 15 minutes counts as late, a more relaxed figure than any other nationality. British managers, on the other hand, agreed that 8 minutes was too much. Collett, who also found no difference from country to country in how often people were late, concluded: "Attitudes to time which are supposedly linked are actually independent of each other. The revealing finding is that cross-cultural differences in lateness are based, not on how frequently people are late, but on how late they actually are."

STILL, THERE ARE SOME SIGNIFICANT cultural differences in time sense, and failing to recognize or adapt to them when one travels from a monochronic culture to a polychronic one, or vice versa, can create what one expert calls "time-related misunderstandings." According to psychologist Robert Levine of the California State University at Fresno, one group who has shown a proficiency for "temporal flexibility" is the thousands of Mexicans who commute daily from Tijuana to jobs on the California side of the border. The comments of one worker who had made the trek for five years were typical. "Each time he crossed the border, it felt like a button was pushed inside him," Levine recounted. "He says that when entering the U.S.,

he felt his whole being switch to rapid clock-time mode. He would walk faster, drive faster, talk faster, meet deadlines. When returning home, his body would relax and slow the moment he saw the Mexican customs agent. He said, 'There is a large group of people like me who move back and forth between the times.'"

Achieving what Levine calls "multi-temporal proficiency" is not easy, but it can be taught. Indeed, time-teaching exercises can prepare travelers and expatriates to deal with faster and slower cultures. In Israel, for example, children from developing cultures are trained to adapt to the country's mainstream pace of life by being taught to value time and efficiency and to appreciate that failure to master this new time sense may lead to larger failures.

When trying to understand the "temporal logic of slower cultures," Levine says it's also important to ask whether people are on "clock time" or "event time." Those on the former, for instance, use the hour on the timepiece to schedule activities; those who follow "event time" allow activities to transpire according to their own spontaneous schedule. Shifting between the two can be difficult, since a move from clock time to event time requires a complete shift of consciousness. As Levine explains, "It entails a suspension of industrialized society's temporal golden rule, that time is money. For those who have been socialized under this formula, the shift requires a considerable leap."

No doubt about it, we are all in the grip of time, whether it be clock or event driven. And perhaps the answer to our opening question, "Does a watched pot boil?" can best be answered by first asking ourselves, the watchers, whether we are working on clock time or event time. The twain may not meet, but the pot will still boil, watched or not...sooner or later.

Y2K SPARKS a blaze of fireworks in Paris. A cardinal point in time, the year 2000 began with much fanfare but little else to mark it. Ultimately no more than a human attempt to measure time, Y2K held only as much reality as any other calendrical moment. Indeed, time's past, present, and future may well be, as Albert Einstein put it, simply a persistent illusion. If so, our clocks and calendars, bells and fireworks are about as valuable as a camera trying to capture a mirage.

CHAPTER 6

DEEP TIME

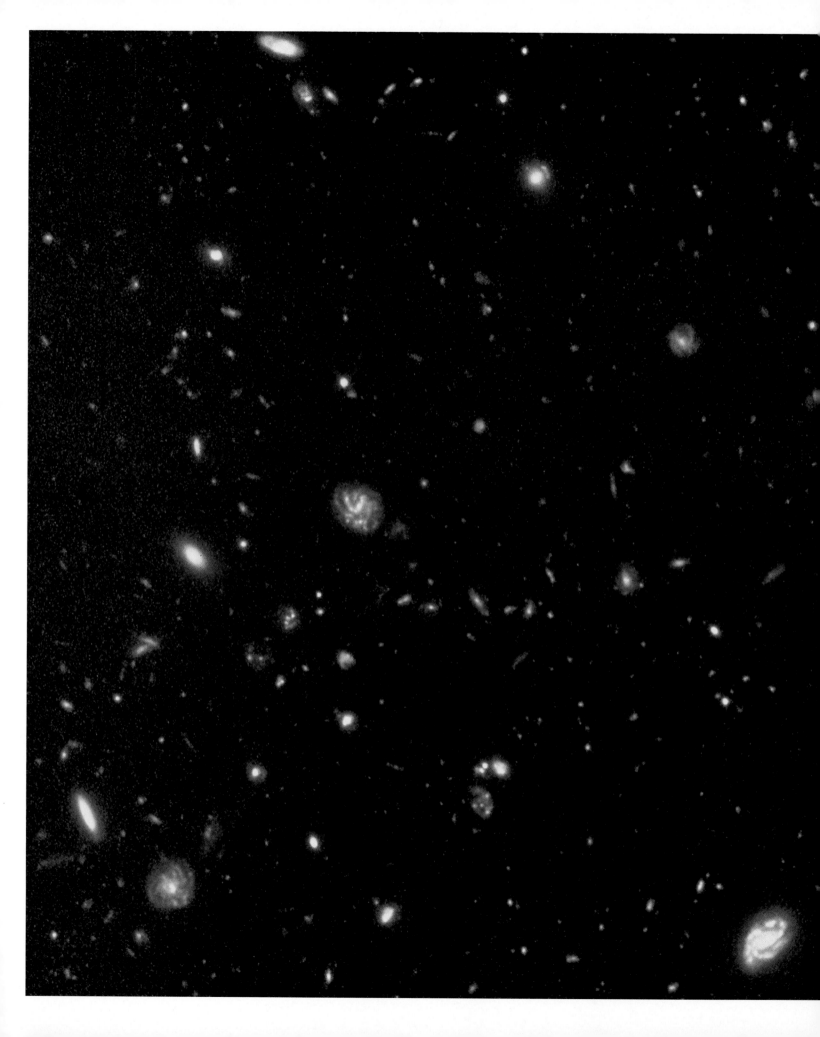

> "Perhaps the biggest surprise of…the past decade is that it is not obvious that the laws of physics forbid time travel. There are many other types of theoretical solutions in general relativity that would allow time travel to the past."
>
> —WILLIAM A. HISCOCK, PHYSICIST

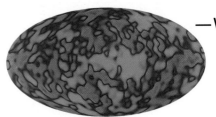

CALL IT WHAT YOU WILL—linear or cyclical, monochronic or polychronic, clock- or event-driven—our sense of time gives pattern and varying degrees of order to what otherwise would be chaos. Whether we see time as a line that stretches out straight as an arrow from here to there and does not repeat itself, or as a circle—which is after all only a curved line—our metaphoric view of time gives tempo to life; without some sense of it, there would be no place in which to insert and space events and experiences nor any way to determine how fast time is "passing." We would be unable to feel the present or describe the past, determine how long it takes for the pot to boil, or make the mental connections that let us appreciate the sequence of notes that form music or the frames that make a movie. We would not know the difference between "says" and "said" when someone speaks. Without a sense of or a way to measure time, nonsimultaneous events would not occur in a definite order with a lapse between them. No question about it: Time may not be a physical constant, but it certainly is with us, constantly flexing its relative muscles.

It is both the invisible glue that keeps us from flying off in all directions and the medium through which everything that moves must pass—and which itself seems to "pass" as things go by. But time is not in fact mathematically segmented, as our clocks and calendars would have it. An abstraction without spatial dimensions, it is nonetheless inextricably linked to the three dimensions of ordinary space in an interdependent, four-dimensional space-time continuum. No object or person, no event or place or action can be located in space and not also in time; nor can anything have a presence in time without also having a presence in space. If we want to establish the distance, or interval, between two events, we need the combination of space and time to do it. Space-time, therefore, is a central concept in the theory of relativity, replacing the old idea of space and time as individual entities.

There comes a point, then, when science takes over from the competing cultural and religious metaphors of cyclical and linear time; when time acquires an actual beginning, a shape, a place in space, and a direction—all based not only on our subjective understanding of time but also on laws that deal with the dynamics of heat, nuclear and electromagnetic forces, gravity, motion, and the mass of material—and on where the very universe is heading.

This time of the astronomer, the physicist, and the mathematician is no more "solid" than it was for ancient philosophers, who noted time's invisible, fleeting presence. But it is more apprehensible. Scientists still haven't seen time with an electron microscope or the Hubble Space Telescope. Nor have they taken it "by the forelock," as Jonathan Swift advised. But they have glimpsed its past, and indeed its birth, in traces of dim and ancient light from far distant stars and from the big bang of creation. They have broken its echoes into vibrating increments tiny enough to be measured by clocks so accurate they will neither gain nor lose a second in millions of years. They have even visualized it, in the face of all ordinary reason, running backward deep in an atom's interior. No longer is time as Newton saw it—an absolute entity between events, the same no matter who measures it.

It is all a matter of one's point of view. For a person traveling in a spaceship at 60 percent the speed of light, a minute is a minute. However, from the point of view of an earthbound observer, time on the ship appears to be running very slowly compared to that on a stationary clock. The two clocks would agree only if they were at rest with respect to one another. "Every reference body has its own particular time," Albert Einstein said. Thus, Einsteinian clocks in motion and in the throes of time dilation seem to operate more slowly than those at rest relative to any observer. Hence, time is personal and relativistic rather than something ticked out by an absolute Newtonian clock.

Some 15 years after his special theory of relativity, Einstein expanded it and added gravity. In his general theory Einstein confirmed that this attracting force of nature does far more, as we now know, than hold a space station in orbit or keep the Earth revolving around the sun and us from floating off into space. Prodigious as such feats may be, they are but a warm-up act for an even more impressive cosmic performance.

GRAVITY TAKES DOWN extraordinary masses of matter, collapsing monstrous stars and, in so doing, making the very fabric of space look like a flimsy canopy onto which a cannonball has been dropped. Under gravity's influence, the geometrical properties of vast regions of space are bent or distorted as huge chunks of matter intrude; and space's configuration is changed. So, too, is that of its cohort, time. In a word, the distortion is best described as curved. Because of changes in the strength of the gravitational field—changes produced by mass or energy situated in various locations in space and in a particular instant in time—space-time is, therefore, curved. The curvature may be localized, but, on a much broader scale, it is also characteristic of the universe.

Because gravity is behind spatial manipulations, and because gravitational fields are an accelerating force, time, which is part of space, is necessarily affected, just as it is by motion. Einstein observed that gravitational fields affect the measurement of time and distance and are capable of slowing down the clock, just as motion does. Clocks close to a massive object, for example, appear to run slower: On the planet Jupiter, a cesium clock would run much slower than one here on Earth. Why is this? First, remember that Einstein said there is no difference between the acceleration produced by gravity and other kinds of acceleration. So, if a moving clock ticks more slowly as its speed of motion gets closer to the speed of light, then the acceleration produced by gravity would also slow time down.

PHYSICIST STEPHEN HAWKING, who holds the same chair at Cambridge University that Newton once did and whose views of the universe have done much to clarify the shape of time, reminds us that "to someone high up, it would appear that everything down below was taking longer to happen." This was tested in 1962 with a pair of hyperaccurate clocks mounted at the top and bottom of a water tower: The clock nearer Earth ran slower, in exact agreement with general relativity. "The difference in the speed of clocks at different heights above the earth is now of considerable importance with the advent of very accurate navigation systems based on signals from satellites," Hawking has observed. "If one ignored the predictions of general relativity, the position that one calculated would be wrong by several miles! Newton's laws of motion put an end to the idea of absolute position in space. The theory of relativity gets rid of absolute time."

Accepting time as relative and settling it into a four-dimensional, space-time continuum still doesn't answer a big question about time: What is its direction? We spoke in the last chapter of an "arrow of time" that seemingly, and logically, indicates a forward movement. But since time really doesn't "go" anywhere, the best way to grasp the arrow imagery is to equate it with a stationary arrow that points in just one direction but doesn't move toward it, like the arrow on a traffic sign or a weather vane. That raises another question: What's behind this directionality, this extension, this reaching out, which, if not indicative of a literal flow, still gives us the impression that time does pass, that it moves forward?

In a word, thermodynamics. It may seem strange that something as insubstantial and illusory as time would have a connection to a branch of physics that deals with the relationship between heat and energy and the conversion of one into the other. Indeed, the four laws of thermodynamics—a 19th-century development arising from studies of steam and other heat engines that convert heat into mechanical energy—make no specific mention of time's direction. However, the second law is, for our purposes, the time-conscious one. It states that in isolated systems heat flows spontaneously only from hot to cold and never from cold to hot. Though the law may seem more in the realm of the physical sciences, it has a connection to the direction of time, notably in the implication that nature—unless there is some intervention—tends toward an irreversible process of disorganization, not organization. Which means that natural events generally go from order to disorder. A word of caution here. In an unfortunate twist of terms, scientists refer to order as disequilibrium and to disorder as equilibrium. A state of maximum disorder is called thermodynamic equilibrium. In nature's wisdom, the preferred condition is for all things to be distributed evenly, or randomly, in what seems to us an unorganized fashion.

Be all that as it may, the irreversible tendency toward what we'll call disorder has been referred to as a thermodynamic arrow of time. So, for all the weight time carries in the four-dimensional continuum, and for all its power to organize and order our lives, it appears to be associated not with order but, paradoxically, with disorder. Moreover, although time and space are interrelated, the thermodynamic arrow provides a sense of direction for time that space lacks. Space is full of direction. We can, for instance, travel forward or sideways between two points in space and not have the feeling that we can't go backward or from one side to the other. We don't get the same feeling when we go somewhere in time, which is why we plan to go places tomorrow but not yesterday.

Using the order-chaos arrow, we see that a tea cup falls from a table and breaks when it hits the floor but doesn't "unbreak" and fall back up to the table; that newspapers go *Continued on page 226*

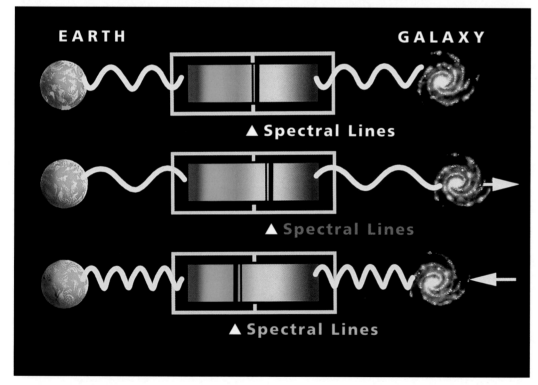

EARTH

GALAXY

▲ Spectral Lines

▲ Spectral Lines

▲ Spectral Lines

THE DOPPLER EFFECT, a phenomenon that applies to all types of waves, can alter our perception of sound and light. As the source of the waves and the observer move relative to each other, the emitted wavelengths appear to change in frequency. The blurred railroad tracks at left suggest a train coming or going; in either case the noise of the passing train will change its pitch, owing to the Doppler effect. Astronomers use the effect to determine whether a stellar object is moving toward or away from Earth. If a galaxy remains at a constant distance from us, the black spectral lines (diagram, top) are "standard"; moving away, the lines are red-shifted (middle); and moving toward us, the lines are blueshifted (bottom). At Jaipur Astronomical Observatory (right) in India, gemstones are used as whimsical representations of the color spectrum.

PROBER OF THE HEAVENS, the Gemini North Telescope on Hawaii's Mauna Kea leans like the Tower of Pisa as it fixes on Jupiter and Venus through a gigantic slit in the dome. The massive instrument provides new views of the cosmos and answers questions that the ancients, with their limited telescopes and time-consuming calculations, could only guess at. By focusing on the rubble of the big bang, astronomers and their machine- and computer-driven telescopes now have a window into the long ago and a means to calculate the rate at which the universe is expanding. While they can look backward in time at cosmic structures that may no longer exist, scientists have yet to know the universe's future, or why the four-dimensional reality of space-time exists at all.

from white to yellow, not the other way around; that rotted fruit doesn't return to ripe; and that snowmen melt but don't "unmelt" to re-form snowmen. One often cited example of how real systems possess a thermodynamic arrow of time is in the diffusion of perfume vapor from an open bottle. As the perfume evaporates, its highly ordered molecules are spread uniformly throughout the room, to the equilibrium state, never to be retrieved. It is the same thing that happens when heat, a form of energy, becomes cold, which is an absorber and a disrupter of heat. Equilibrium, or disorder, is thus brought on by the distribution of heat at a uniform temperature.

When given their own head, all things, then, are caught in a process of increasing natural disorganization, not organization. As things decay, break apart, and run down, their original state is gone for good. In the end, they are in a state of thermodynamic equilibrium, maximum disorder that is arrived at in the future, not in the unalterable past.

It is this tendency toward disorder—the measurement of which is referred to as entropy—that increases with the "forward" direction of time and that, like the psychological arrow, reinforces our notion of a one-directional arrow of time. But the disorder that grows out of order is not dictated by any formal "directionality" stamped into the underlying laws; it is simply because there are more possible disordered states in nature than ordered ones. Hawking uses a jigsaw puzzle to demonstrate this. "Suppose the pieces of the jigsaw start off in a box in the ordered arrangement in which they form a picture. If you shake the box, the pieces will take up another arrangement. This will probably be a disordered arrangement in which the pieces don't form a proper picture, simply because there are so many more disorganized arrangements."

There is yet another arrow associated with a movement toward disorder and with the idea that time has only one direction, forward. It is a cosmological one that directs our telescopes' eyes to the expanding universe. If time began with a big bang many billions of years ago, as many scientists believe, the universe is moving outward, propelled along on a never ending path by the force from a cataclysmic blast of incredibly dense matter compressed into a tiny sphere. As space expands, gaseous debris condenses, spawning stars, galaxies, planets, and, in some fortunate zones of the increasingly disordered vastness, the seeming paradox of complex, ordered, and intelligent life. Under constant acceleration and showing no signs of falling back on itself—although a few theorists suggest an eventual shrinkage back to a "cosmic egg" that will explode again to create a new universe—the current universe is on a one-way track to the future, not to the past. (Some scientists believe that if the universe is an isolated system, all its energy will eventually be reduced to a uniform heat and distributed widely, resulting in a condition of final and enormous disorder, dubbed by 19th-century scientists "cosmic heat death.")

THORNY QUESTIONS ARISE when we consider the incredible complexity and order of life. Isn't the growth of an infant into an intelligent adult evidence of another life process, one that causes entropy to decrease not increase? Besides, if the second law is right, how could we have arrived at our current ordered state if we, and the universe, grew out of something less complicated than ourselves, or out of something as seemingly chaotic and disordered as the big bang? (Most scientists don't characterize the bang as disordered, since it was the ultimate unification of all things.) Aren't advocates of evolution violating the second law if they argue that beings as wondrous as us sprang from a jumbled soup of primordial chemicals, or from the most rudimentary of organisms?

Insofar as the big bang is concerned, nobody really knows what the state of the universe was at its creation. Hawking tells us that it could have started in a very smooth and ordered condition, which would have led to our well-defined arrows of time. On the other hand, it could also have begun lumpy and disordered, in which case the universe would already be mired in complete disorder, so that disorder could not increase with time. It would either remain constant, with no well-defined thermodynamic arrow of time, or the disorder would decrease, turning the arrow in the opposite direction—a cosmological arrow that points in the direction of an expanding universe. Neither of the last two possibilities, says Hawking, agree with what we observe. Moreover, when all else fails, we might suppose, as Hawking does, that God decided the universe should finish up in a state of high order "but that it didn't matter what state it started in."

When it comes to human life and how it does or does not tie into entropy, things get even murkier. We know, for instance, that the second law of thermodynamics works only in an isolated system, that is, one that can exchange neither matter nor energy with its surroundings. If it works as it should in the confines of human beings, the growing process should involve a move from order to disorder. But given the intricacy of a human being, one can argue convincingly that this seems not to be the case and the second law doesn't apply. Yet it does. First of all, the growth process is not going on in an isolated system (the body), where order not only goes to disorder but the probabilities of disorder are enormous. It goes on because there is an entropy increase somewhere else—in the air, food, heat, light, and water that enter the system. True, if we consider the human body as a mere container of skin that seals in inert biochemicals and useless organs, then we may have a case. But there is increased entropy associated with all the biochemical processes required to put human beings together. It's something like what happens when water freezes, a process that makes us think disorganized water has wound up as an organized lump. Entropy decreased as the water hardened, but there was an increase in entropy in the air where the heat from the water went in the first place. "A strong thermodynamic arrow is necessary for intelligent life to operate," says Hawking. "In order to survive, human beings have to consume food, which is an ordered form of energy, and convert it to heat, which is a disordered form of energy." Entropy, thus, is more than just a tea cup smashing permanently to bits on a floor.

THE UNIVERSE'S CONFIGURATION, anomalies, and behavior also affect time's status and direction. Consider the ultimate result of the second law of thermodynamics: A black hole, an invisible region in space-time from which nothing, not even light or radiation of any kind, can escape because gravity there is so strong. Formed when massive, dying stars collapse, it is an as yet unknown state into which anything or anyone falling would, in Hawking's words, "soon reach the region of infinite density and the end of time." Losing time here is literal, for what goes in is lost, probably irreversibly, through the gravitational disorder gained by the universe as it develops.

A multitude of other strange things occur in such an intense gravitational field. If a person got inside the event horizon, or boundary, of a black hole and survived—a highly unlikely possibility— he or she would be stretched into a thin band by forces similar to but fiercer than those responsible for the Earth's tides. Our unfortunate traveler, who would ultimately be torn apart, would at first see objects outside the hole as distorted, because gravity bends light. He or she would not be seen from outside, however, since light cannot escape past the event horizon. Nor could the traveler emerge, since the escape velocity required for matter to free itself from such a crushing force would have to exceed the speed of light, an impossibility.

Time is also affected, relatively speaking, as the traveler is drawn deeper into the hole: From the traveler's point of view, the trip to the center of the dead, collapsed star would take far less than a split second; from the point of view of an outside observer, however, the fall takes an infinite time and is virtually frozen.

Black holes can't be seen directly because light can't get out of them, but their presence can be deduced by the effects of their intense gravitational pull on nearby matter. Neither can we "see" the universe directly or where it is going. But by combining observations of the microwave background radiation coming to Earth from deep space with the "fossil" light from distant stars with the supercomputer models of great aggregations of galaxies, scientists can tell us much more about the structure of the universe, from its creation to the present.

Microwave radiation is especially telling. It is the dim remnant of the radiation released in the big bang's enormous blast of heat—an echo, a snapshot of that event, still reflecting the structure of the universe about 300,000 years afterward. Incredibily uniform, it has the same intensity in every direction scientists look in space, about one part in 10,000. Along with the other observations, background radiation has given us a history of the universe dating back to an astonishing ten million-trillion-trillion-trillionths of a second after the big bang. At that time, called Planck time (after the German physicist Max Planck, one of those responsible for quantum theory), gravity presumably separated from the other fundamental forces of electromagnetism and the so-called weak and strong forces that work at the subatomic level. Particles, nuclei, and eventually atoms

formed, followed by clouds of hydrogen and helium that would condense into stars and galaxies.

Through the entire convulsive process—from the primordial fireball through the cooling down period—time, which was born in one instant along with matter and energy—has been leaving its space tracks. Armed with the knowledge that light travels at one fixed speed, astronomers can look at the night sky and determine the distance to the farthest object in their view. The light from a star ten million light-years away began its journey ten million years before it reached the observers' eyes; that from another one "nearby" at five million light-years distance began its journey five million years before the astronomers spotted it. Seen at exactly the same time, the two twinklers may look the same, but they differ enormously in age. Indeed, we may see it as it was when the light left its surface at a speed of more than 186,000 miles per second. Or look at the sun and you see it as it was eight minutes ago; a blast on the moon would be visible two seconds after it occurred.

Scientists don't know what existed before Planck time, nor can they rely on time, even though it points forward, to predict the universe's future behavior. Nonetheless, theories about cosmic destiny are plentiful: The universe is open, expanding for eternity and toward infinity; it is closed but still expands until gravity takes hold and causes it to collapse; it collapses and rises again and again in a series of big bangs spaced over eons; it is in a "steady state," a place with no beginning or end, where the density of matter is constant, where new galaxies continually replace old ones.

Hawking has even proposed a "no boundary" universe, one of expansions and contractions, finite but with no boundaries. In such a place there are well-defined arrows of time, but even if the universe re-collapses, the thermodynamic and psychological arrows would not reverse, according to his theory,

GANGLIA OF COLORED WIRING wraps a particle detector in Hamburg, Germany. Used to penetrate the bizarre subatomic world of the infinitely small, these macroscopic instruments attempt to measure microseconds of time, though subatomic time may run backward—or not exist at all.

FOLLOWING PAGES IN A "STREAMER CHAMBER" at CERN, laboratories, near Geneva, Switzerland, charged subatomic particles gush out in a whirl of microcosmic beauty. In re-creating the state of the universe microseconds after the big bang, physicists may gain insights into the origins of time.

nor would disorder decrease. Arrows that tell us the difference between past and future are one thing, but the classical laws of physics, at least, apparently do not distinguish between any forward or backward directions of time. As far as we know, broken eggs will not reassemble and fall up onto the table when the universe is collapsing.

If, as most scientists are convinced, time "moves" in one direction (and we all know what that is), it would appear that any idea of backward-running time is not much more than fodder for fiction. Well, maybe yes, maybe no. When we take time to the subatomic level, the arrow of time seems to behave quite differently than its counterpart in a compass needle, with its consistent northward direction. For in the bizarre quantum world of quantum mechanics—the branch of physics that deals with the interaction of matter and radiation, with atomic structure and the motion of ultrasmall particles— laws cannot be explained by experimenting with more hefty everyday things. (A quantum is actually the smallest amount of any physical quantity that can exist independently, like, say an individually distinct quantity of electromagnetic energy.) "Things on a very small scale behave like nothing that you have any direct experience about," said physicist Richard Feynman. "They do not behave like waves, they do not behave like particles, they do not behave like clouds, or billiard balls, or weights on springs, or anything that you have ever seen."

So how does time differ in this spooky world of the subatomic? Well, if we consider the shifting behavior of some subatomic particles, one can make a case that some do indeed *Continued on page 236*

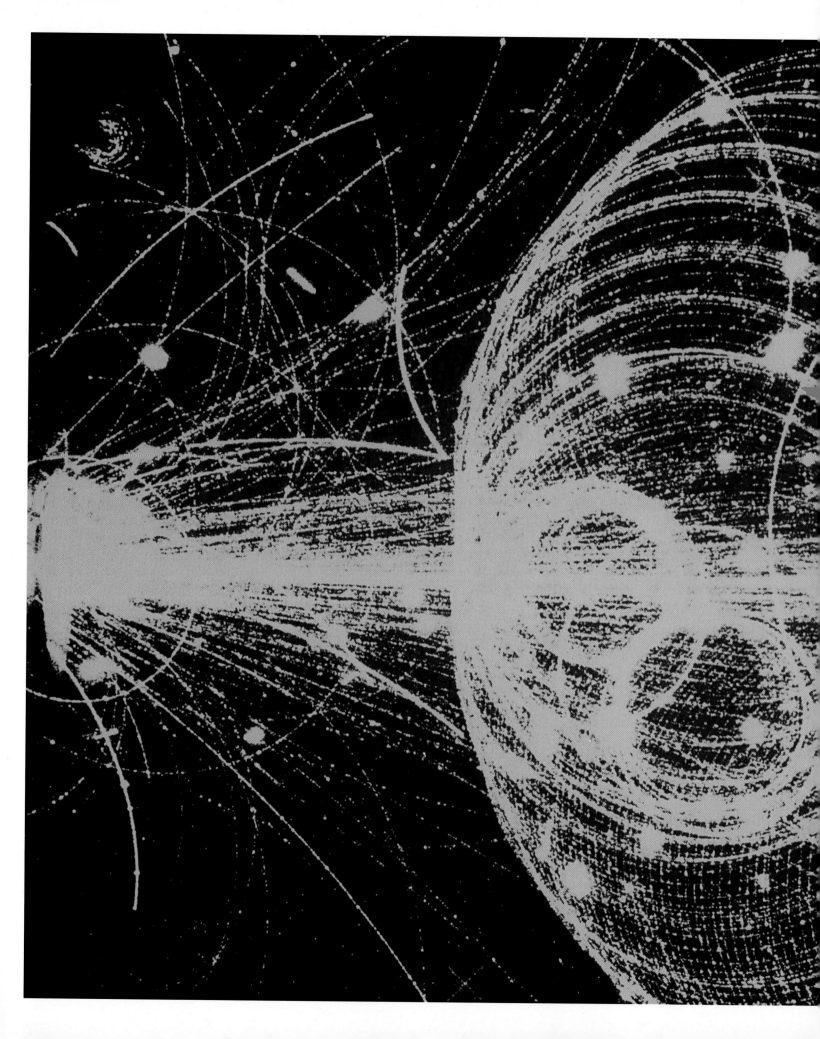

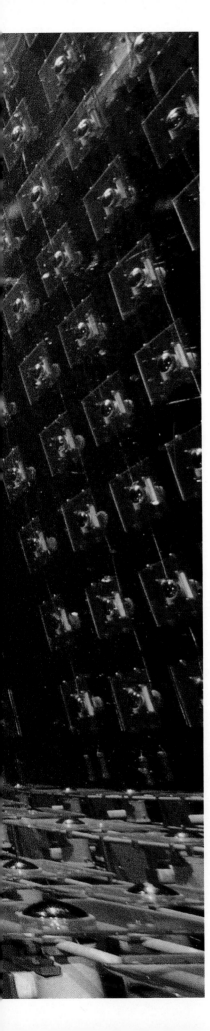

MESSAGES FROM THE PAST: Some 2,000 feet beneath Lake Erie, a diver (left) prowls a pitch-dark tank of water designed to detect elusive neutrinos released from exploded stars. Uncharged subatomic particles with little or no mass, neutrinos travel at the speed of light and are extremely difficult to find. The 2,048 light sensors lining this tank managed to detect sparks from eight neutrinos as they ran vertically through the water and interacted with other particles. Perhaps 100 billion neutrinos per square inch can strike the Earth when a supernova collapses. To hunt them, scientists have also spread neutrino-detectors on the ocean floor and underground in salt mines. The image at right, captured by the Hubble Space Telescope focusing 8,000 light-years into the cosmos, shows the Etched Hourglass Nebula, a shell of gas expanding.

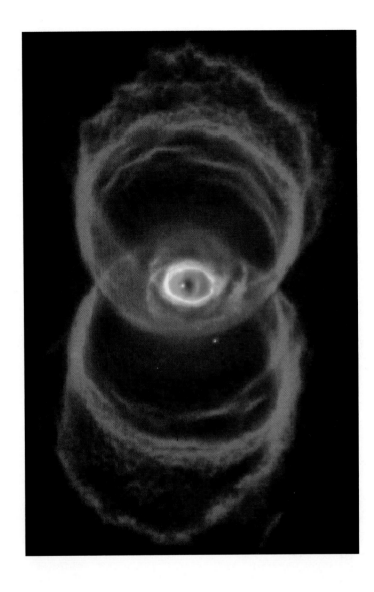

PAST BECOMES PRESENT in space-time, a continuum of the three spatial dimensions plus a fourth—the relative, changeable dimension we know as time. Distance and the speed of light become key to understanding this illusive dimension. Because light always moves at 186,282 miles per second, we can measure how far from us a heavenly body is and how remote it is in time. In the illustration at right, space-time becomes a cone of light, with panels representing space at different moments; the tip of the cone may be any point —Earth, for example— at that instant. Events, thus, can be divided into those whose light has had time to reach us, and those that are still unknown to us. As Earth moves through time from left to right, our cone spreads on each panel, sweeping events into our "present." We cannot see a star explode (top of middle panel) until thousands of years after the event, when its light finally reaches us (right panel). Time's relative nature is best imagined through moving objects, like a train (opposite): If the train were hurtling along at close to the speed of light, time on its onboard clock would drag compared to the time recorded by a stationary clock.

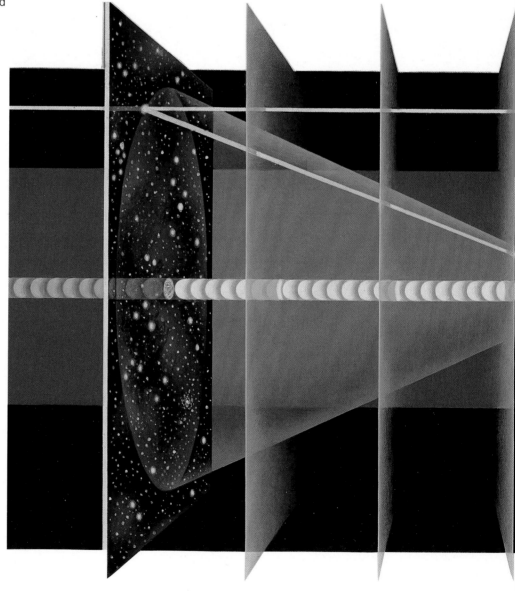

seem to run backward in time, while others prefer going conventionally forward. This isn't exactly easy to witness. It's not like watching a filmed racehorse galloping backward when the film is reversed. In that case, we know and can see that the racehorse doesn't actually run that way. What goes on at the subatomic level is more difficult to grasp. First, physicists long believed that the essence of reactions between energy and matter doesn't change—that is, they'd work exactly the same if left were exchanged for right. But in recent years, physicists have discovered that nature *does* seem to prefer certain kinds of "right-handed" reactions over "left-handed" ones.

Evidence of this emerged recently in experiments conducted at CERN, the European Laboratory for Particle Physics in Geneva. There, researchers collided atoms to create particles called kaons and their antimatter counterparts, antikaons.

(Antimatter, which differs from the matter that predominates in our part of the universe, is composed of oppositely charged antiparticles.) As the particles moved about, kaons transformed into antikaons, and vice versa, all under the eyes of a detector that counted the particles as they decayed. The result: The rate for antikaons that turned into kaons was higher than in the time-reversed process of kaons becoming antikaons. The kaon behavior, according to the researchers, essentially followed what the researchers called "a microscopic arrow of time." Whether it was a unique form of direction raising the notion of both forward and backward causation—and the possibility that the future, at least at the fundamental level, may perhaps be able to affect the past—is another matter. But as the CERN scientists put it, "This experiment casts a new and sharper light on the fundamental symmetries of physics and how they

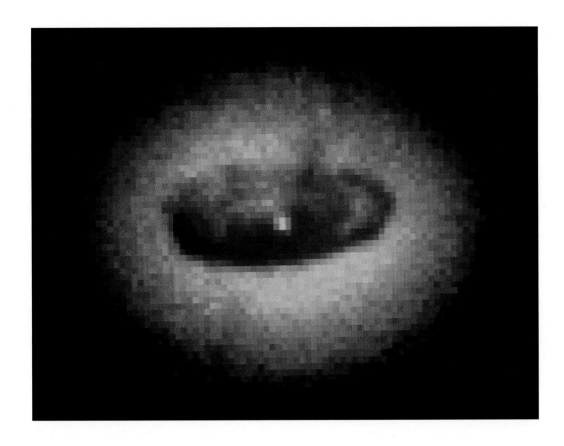

COSMIC DRAIN:
Created by the collapse of a massive star, a disk of gas and dust swirls into what is almost certainly a black hole at the center of galaxy NGC4261. With a diameter of some 800 million light-years and heavy with 500 million times the mass of our sun, the hole creates an intense gravitational pull that prevents even photons of light from escaping. (The bright point at the center is the final flare of heated matter.) Inside the center of a black hole, mass becomes so densely compacted that space, without which there is no time, does not exist.

are broken. Physicists know that time symmetry has to be part of a larger, more powerful package (including time reversal), which sits at the heart of modern physics. Swap antimatter for matter, view the universe in a mirror, and reverse the direction of time, and all experiments should come out the same way they do in the real world."

If time symmetry is indeed violated by the likes of kaons, and if we let our imaginations run wild, we can cite such laboratory manipulations of time to raise several intriguing questions. If different clocks time different times, can we set one to record a cause happening after its effect? Can we alter events that have already happened? Is there really a difference between past and future? Could a space traveler return to his point of origin, to the time before he left? If space-time is curved, and we traveled around the universe, would we come back to where we started? Are there

universes in which time travels backward? Will we ever be able to travel backward and forward in time?

The last question is the one that has captivated just about everyone at one time or another, whether or not they know anything about spatiotemporal events on a microscopic level. Writers of science fiction have, of course, fed the imagination. H. G. Wells gave us some insight into what it might be like to journey out of the present:

"I am afraid I cannot convey the peculiar sensations of time traveling," he wrote in his book, *The Time Machine*. "They are excessively unpleasant. There is a feeling exactly like that one has on a switchback (a zigzag road)—of a helpless headlong motion! I felt the same horrible anticipation, too, of an imminent smash. As I put on speed, night followed day like the flapping of a black wing. The dim suggestion of the laboratory seemed presently to fall

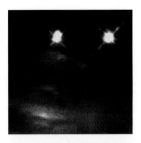
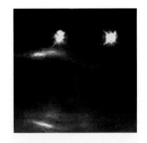
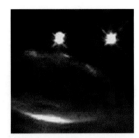

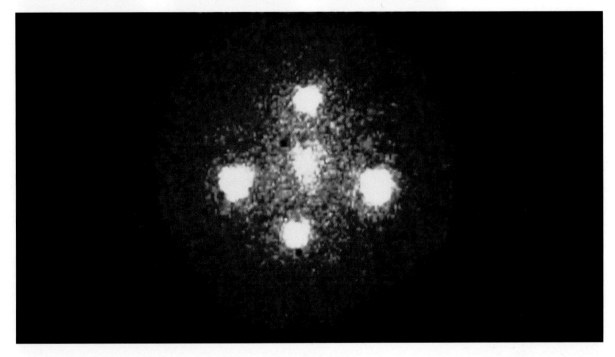

IF CHICAGO'S SEARS TOWER (opposite) were solid rock and could be compressed to the density of a neutron star, it would be no larger than the half-inch-high ball held by University of Chicago professor Dave Arnett. But it would weigh ten billion pounds. Thought to be a rotating neutron star, a pulsar (shown in the images above, taken months apart) can trap high-energy electrons in its magnetic field, giving it a shifting halo effect. "Einstein's cross" (right), taken by the Hubble, shows a closer galaxy surrounded by a more distant quasar—an object larger than a star but smaller than a galaxy. Its image is repeated three times in a mirage-like effect caused by "gravitational lensing," an effect postulated by Einstein in 1936. As the quasar's light passes an intervening galaxy, the galaxy's gravitational pull bends the light, creating the optical illusion.

away from me, and I saw the sun hopping swiftly across the sky, crossing it every minute, and every minute making a day…. The slowest snail that ever crawled dashed by too fast for me. The twinkling succession of darkness and light was excessively painful to the eye. Then in the intermittent darknesses I saw the moon spinning swiftly through her quarters from new to full, and had a faint glimpse of the circling stars. Presently as I went on, still gaining velocity, the palpitation of night and day merged into one continuous gray. The sky took on a wonderful depth of blue, a splendid luminous color like that of early twilight. The jerking sun became a streak of fire, a brilliant arch in space, the moon a

fainter fluctuating band. I could see nothing of the stars, save now and then a brighter circle flickering in the blue."

It might well be that way, but unfortunately (perhaps fortunately, if we consider the possibility of getting stuck somewhere we just wanted to visit) our civilization is not yet advanced enough to build a transport system that could violate time symmetry and, in so doing, deal with all sorts of paradoxes, cause-and-effect distortions, physical and psychological stresses, and inconceivable speed and force.

When H. G. Wells's time traveler returned and noted that his machine was "a little travel-worn" and that the sensations he experienced while breaking

EARTH WITH A VIEW:

The farther out we look, the farther back in time we see. And from the Earth (positioned here as it was at midnight Greenwich mean time, January 1, 2000), we can look out 11 to 15 billion light-years, back to the presumed age of the universe. Distances are not drawn to scale here but increase dramatically as they become more remote. Light from the moon, with a mean distance from the Earth of a mere 238,866 miles, takes but two seconds to reach us. In contrast, light from the galaxy M87 (top center) takes 50 million light-years to reach us, so we see it as it appeared 50 million light-years ago. The outer ring (shown in pink and blue) represents the glow from the microwave background 300,000 years after the big bang occurred, marking the beginning of what we know as time.

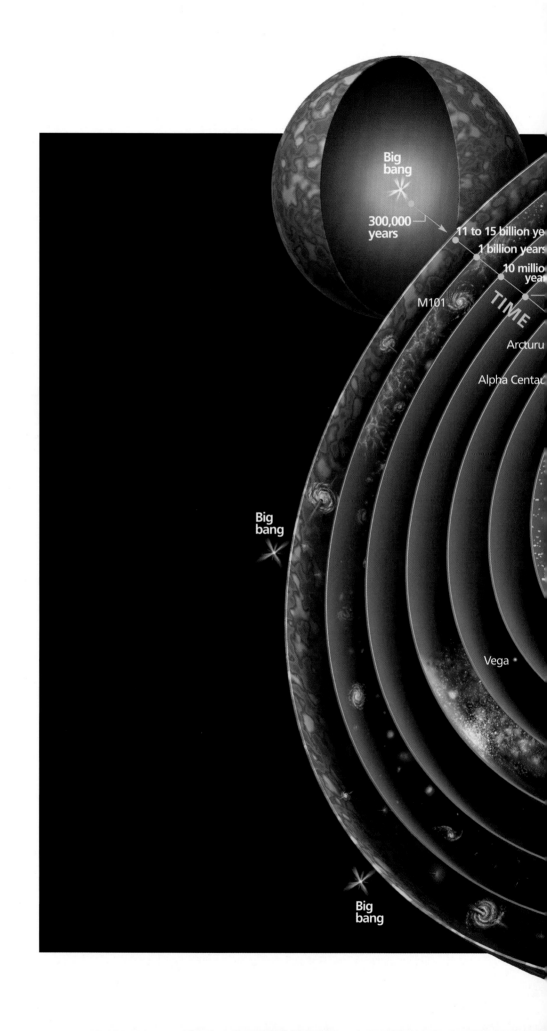

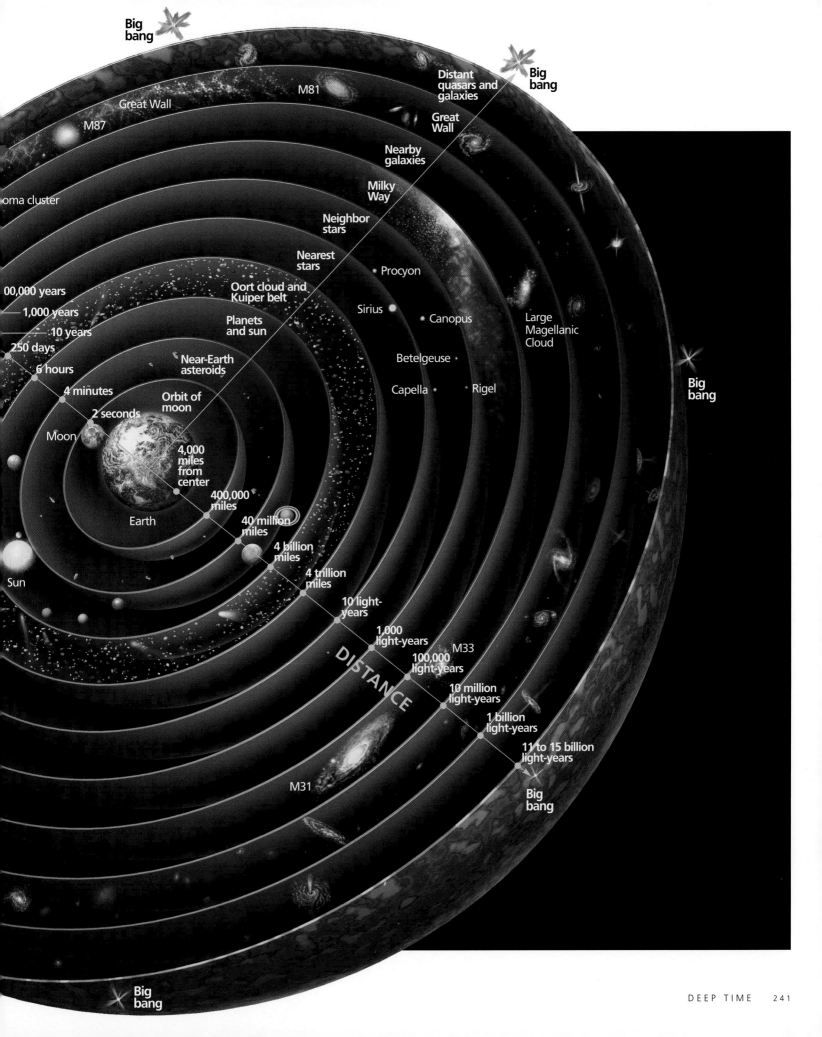

Big bang

M81

Distant quasars and galaxies

Big bang

Great Wall

M87

Great Wall

Nearby galaxies

oma cluster

Milky Way

Neighbor stars

Nearest stars

Procyon

00,000 years

Oort cloud and Kuiper belt

Sirius

Canopus

Large Magellanic Cloud

1,000 years

Planets and sun

10 years

250 days

Near-Earth asteroids

Betelgeuse

6 hours

4 minutes

Orbit of moon

Capella

Rigel

Big bang

2 seconds

Moon

4,000 miles from center

400,000 miles

Earth

40 million miles

Sun

4 billion miles

4 trillion miles

10 light-years

1,000 light-years

100,000 light-years

DISTANCE

M33

10 million light-years

1 billion light-years

11 to 15 billion light-years

M31

Big bang

Big bang

temporal bonds were "excessively unpleasant," he didn't know the half of it. Some of the bothersome questions he might have raised before his trip, and perhaps answered after his return, have baffled scientists, philosophers, and theologians for centuries. If you could go back to a time before you were born, what would happen to your memories? Would they be of the future? If you reached yourself as an infant, could you kill yourself? If so, how did you get there to begin with? If you're always hostage to your own present, how can you travel ahead to your own future, to the not yet? On the other hand, could you travel to someone else's future, since you're not living in his or her present, or to the world's future, and if so, how do you get to their time lines? If you could travel back in your own past, would it include regressing through your inherited genes, allowing you to switch chromosomes somewhere and become your own father or mother? Could you get back fast enough from a trip so that you could pick up where you left off, or would everyone you know be long dead. Does time pass when you're gone? Finally (not really, since there are many more questions), if time travel has been developed by far-advanced civilizations, where are all the travelers and their machines? Should we suspect that they're here already whenever we learn of someone making a killing at a lottery drawing, on the stock market, or at the racetrack—each by "predicting," or knowing, the future.

It's all enough to make one appreciate that our memories are the easiest and safest means of time travel, and that gazing deep into space is the best look we'll probably ever get of the past. Our questions may seem illogical and fanciful, but we can blame time for the curiosity behind them. For it is time that raised the issue of its direction in the first place, with its arrows of the mind and of thermodynamics and its crazy quantum reversals.

And so instead of automatically ruling out time

GRAVITY—ANTIGRAVITY
SHAPERS OF TIME AND SPACE

Falling through a vacuum, a feather and an apple fulfill Galileo's bold 1638 prediction that all objects drop at the same rate of acceleration. To test this, he dropped objects of different weights from a height—legend says it was from the Leaning Tower of Pisa—and observed that they reached the ground simultaneously. Like time, gravity is 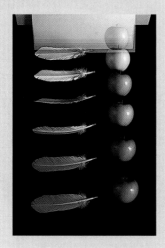 impalpable, but its effects, like those of time's passage, are clearly visible.

Although it is the weakest of the four basic forces in nature, gravity is still capable of distorting space, as it does when it creates a black hole. And it can bend time and slow clocks. But whether gravity is some sort of universal force of attraction or is composed of hypothetical particles called gravitons, no one can truly say. Nor can scientists fathom what its bizarre antithesis might be. Antigravity, an invisible, putative force that seems to repel rather than attract, is believed by some scientists to permeate "empty" space as "dark energy." Along with the equally mysterious "dark matter," which may well make up most of the matter in the universe, antigravity energy may be the force that counteracts gravity. Instead of attracting cosmic objects to one another like conventional gravity, antigravity may serve as a repellent, filling the space between galaxies, and in effect expanding the "emptiness" that separates them. If this is so, cosmic expansion is speeding up as antigravity strengthens and increases over distance, and ordinary gravity, assumed to be slowing the outward movement, weakens and loses its grip on the cosmos.

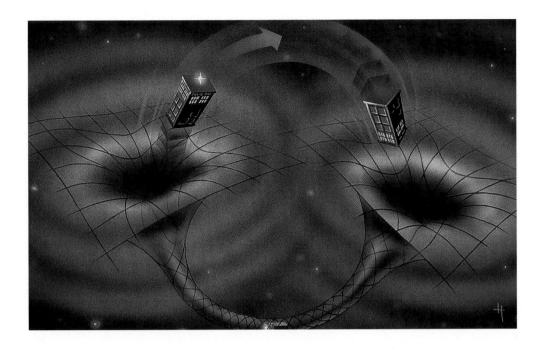

SHORTCUTS BETWEEN PAST AND FUTURE, the theoretical space-time links called wormholes could connect two black holes within our universe— or even black holes in two different universes. Whether wormholes exist at all is questionable, and, if they do, could they be used for space travel? Are they large enough to handle a spaceship—or a microscope? And if a spaceship could navigate a wormhole, would it need to travel faster than light, because the hole, under the enormous influence of gravity, wouldn't stay open long enough for even light to travel from one universe to another? Finally, would a black hole's enormous gravitational pull rip apart an approaching spaceship before it even got to the wormhole?

travel as preposterous, the uncertainties stimulate theorizing, and several physicists and mathematicians, along with an assortment of amateur "mathemagicians," have begun to take it seriously. Their conclusion is that, in principle at least, the laws of physics and of relativity might not forbid sojourns through time. Indeed, there's enough disagreement over what relativity does and does not forbid that virtually all speculation about time travel seems logical and sometimes even scientifically acceptable.

But how to do it? First of all, we can forget about a "machine" like the one with a riding saddle and starting and stopping levers that H. G. Wells's time traveler cobbled together out of brass rails, ivory and nickel bars, and a quartz rod. A spaceship might do the trick (the universe is a huge place, after all, and there's a lot of time in all that space), but for its passengers to get to, say, the future in a reasonable lifetime, the ship has to be able to travel at close to the speed of light. At that speed, says the time dilation dictum of special relativity, time would slow markedly for the traveler relative to time on Earth. It

would then be a matter of practicing patience: According to William Hiscock, a physics professor at Montana State University, a round-trip to the center of our galaxy and back to Earth, a distance of some 60,000 light-years, could be completed in a little more than 40 years of shipboard time. When the space travelers returned to Earth after "existing" during all the times along the way, they'd be 40 years older, but 60,000 years would have passed on Earth.

Sounds like a piece of cake. But even if a spaceship capable of such inconceivable speed could be built, there is another formidable obstacle: finding a propulsion system. Relativity says that the mass of a moving object increases with its velocity and that as the velocity approaches the speed of light, the ultimate limit of speed, the mass nears infinity. At the speed of light, mass would become infinite. This means that enormous force would be required to accelerate an object flying even near the speed of light; an infinite force would be necessary to accelerate an infinite mass. The trip to the future described by Hiscock would require an amount of energy

UNDULATING DOWN AN EXPERIMENTAL TUNNEL at the Laser Interferometer Gravitational-Wave Observatory (LIGO) near Livingston, Lousiana, a flow of laser light (shown in this time exposure) is designed to detect ripples in space-time. Called gravitational waves and predicted by Einstein in his general theory of relativity, the ripples are created by violent events in the far-distant universe. Both this LIGO facility and a second one in Richland, Washington, hope simultaneously to detect such an event. Although this has yet to happen, the influence of such waves on neutron stars orbiting each other has been measured accurately. LIGO findings could not only unravel some of the mysteries of gravity but also give astronomers a new window into the universe itself.

greater than a planetary mass. Putting this in terms of current technology, Marc Millis of NASA's Space Propulsion Technology Division explains that, using today's space rockets, a journey to the nearest star beyond the sun would burn far more fuel than the entire weight of the known universe—and it would still take nearly a thousand years to get there. "If you use a nuclear fission rocket, you need about a billion supertankers of propellant. If you use a nuclear fusion rocket, you only need about a thousand supertankers. And if you assume that you'll have a super-duper ion or antimatter rocket, you'll now only need about ten railway tankers. It gets even worse if you want to get there sooner."

So, FOR A DETERMINED TIME TRAVELER is there a way out ? There may be, if we consider such oddities as Star Trek-like warps, wormholes in space to rocket through, and ships that operate under the strange time-reversal qualities of subatomic particles and are powered by antimatter. In theory, at least, such avenues to the future and the past may be remote and more appropriate to science fiction, but several scientists feel they're worth examining. Speed is still a requirement, enormous speed faster even than the fastest thing we know—the speed of light. Here's where the hypothetical warp drive comes in. Based on the idea of an inflationary universe and the belief that during the first moments of the big bang space-time expanded faster than the speed of light, warp drive is an approach that would circumvent the speed limit imposed on matter by relativity. If space-time can streak along faster than the speed of light, why not a spaceship powered by an engine that would work by moving space around matter? Since space may be nonmaterial, such a ship could travel faster than the speed of light and would be propelled weightlessly along through space-time. "All you need to do," explains Millis, "is contract space-time in front of your ship and expand space-time behind your ship. This 'warped' space and the region within it would propel itself 'with an arbitrarily large speed.' Observers outside this 'warp' would see it move faster than the speed of light. Observers inside this 'warp' would feel no acceleration as they zip along at warp speed."

One analogy used to explain how it all would work is a moving airport walkway. There may be a limit to how fast one can walk across the stationary floor (analogous to the light speed limit), but, when one walks on a moving section of floor that goes along faster than one can walk, the result is akin to a moving chunk of space-time. The expanding space-time behind the ship is analogous to the place where the moving walkway emerges from beneath the floor; contracting it in front of the ship is the equivalent of the walkway running back into the floor.

There are, of course, catches. No one yet knows how to manipulate space-time, how to turn the exotic energy required on and off, or even whether enough of the sort of energy required to distort space-time exists. Finally, there's the question of whether the warp would really beat light speed. For now at least traveling faster than light is still the stuff of fiction, beyond the grasp of current physics and engineering.

Another possible way to time travel would be through a so-called wormhole, a theoretical tunnel-like shortcut between two widely separated points in space. If wormholes exist, they might connect not only regions of our own universe but also a site in our universe to that in another. A black hole situated at one place and time could, conceivably, be connected by a wormhole to another black hole located in another place and time (or in the same place but at a different time). To envision a wormhole, author and astronomy professor Timothy Ferris of the University of California at Berkeley uses the pop-science image of cosmic space-time as a rubber sheet full of the infinitely deep depressions made by black holes. Each has a fluted stem, like that of a bud vase. "Bend the rubber sheet so that the open, bottom ends of two such stems are joined together. The result is a wormhole [which] could provide the ultimate in efficient travel: a wormhole a mile long might connect two regions of space hundreds of light-years apart."

Wormholes, thus, would serve as time machines through which travelers could visit ordinarily inaccessible regions of space, passing through time to, say, the past, the "place" that many people think about when they think of time travel. But where do you find a wormhole? Can you make one, especially one large enough for a human or a spaceship to fly through? And even if we had one, would it work as a time machine?

Quantum speculators suggest that space-time consists of a complex, foamlike structure of wormholes on the smallest scales—a billion billion times smaller than an electron. "Some physicists believe it may be possible to grab one of these truly microscopic wormholes and enlarge it to usable size," says Hiscock, "but at present these ideas are all very hypothetical." Still, some scientists are undaunted. If a subatomic particle may be able to tunnel backward in time, why not fix it so that a human can hitch a ride? Theoretical physicist Chris Van Den Broeck of the Catholic University in Leuven, Belgium, recently came up with an innovative warp-drive idea that seems to be another paradox: Simply inflate the inside of a microscopically small "warp bubble"— which would need only a relatively small amount of energy to get it flying at warp speed—and keep the outside tiny. That way, you'd have a bubble smaller than an atomic nucleus but with an interior large enough to contain a starship! It makes for interesting reading, mathematically speaking, but the calculations, other physicists say, just don't add up.

Some of the other paradoxes that riddle time travel have been explained, but even those explanations require enormous leaps of imagination. For example, take the classic one about traveling back in time and doing away with our grandmother, or with our own infant self, an impossibility for the logically minded. Conventional wisdom, if there is such a thing when it comes to talking about time trips, says that we might be able to travel back, but that we wouldn't be able to change the past; we'd be merely outside observers looking on, just as we see the past in the light from faraway stars. In any event, say the practical nonbelievers, something, we don't know what, will always intervene to prevent us from damaging what nature and destiny assembled.

Another kind of wisdom argues that we can indeed change past events. If there are other universes and they're connected by wormholes, we might pass from one to the other and emerge in a world that is far different from ours—or the same—into what's been called a parallel universe. Here, time travelers might find parallel versions of themselves, but each slightly different. In such universes, all conceivable outcomes might be possible, including those influenced by an intruder from another universe. Fantasy, perhaps, but in a universe where kaons are fouling up time symmetry, and scientists can envision objects that are microscopic on the outside but macroscopic on the inside, anything might be possible.

If we look at time travel as a journey through a space-time that curves back on itself because of gravity, then there is no difference between the direction of time and directions in space; we go forward as we go backward, and presumably we return to where we began. But will space-time always be with us? Will it be there long enough for us to be able to travel in time? Will time end? When?

Allusions to eternity have been with us for as long as humankind has pondered the question, "What is

LIFE ON MARS?
Ultimately, the answer to that will only come from a manned mission. Gearing up for such a mission, NASA has developed a prototype suit (left), engineered for limb and finger flexibility, allowing future astronauts to collect rock samples. Such expeditions are still more than a decade away; in the meantime alien signals and the discovery of worlds like our own continue to elude us. Still, the quest stirs the imagination. If intelligent life does exist somewhere else in the universe, it will probably share some basic characteristics with our own. The composite image of an alien being (right) incorporates physical and chemical properties that are universal—even though its appearance is decidedly out of this world.

time?" A folk riddle asks the question, "How many seconds in eternity?" Eternity, as we conceive of it, is seemingly endlessness, an indefinite, immeasurable expanse of time. It is what the mythic Greek king Sisyphus faced when his punishment was to roll a stone uphill, and, nearing the top, had to watch helplessly as it slipped from his grasp and plunged to the bottom…again and again and again, forever.

Whether time is infinite or finite is not easy to answer. If time needed an event in order to be born and more events to survive and events quit happening, then time as we perceive it would probably end someday. If it lives on after events are gone, then it's infinite, although the absence of any psychological arrow of time in such a bleak nothingness would make that quite difficult to measure and confirm. On the other hand, scientists may

argue that events will always be, if only in the invisible particles that keep moving farther apart in an open-ended, expanding universe.

Perhaps an answer will come when scientists realize their dream of a grand unified "theory of everything," welding together all the fundamental forces, each of which affects and is affected by time one way or the other. Until then—and if then—time remains a mystery about which we can say no more than that it simply began, is now, can be measured, and will probably continue to be. Like life, it will just go on. But even farther beyond our comprehension is why any of it, time and the universe, is there. As Stephen Hawking has said, "If we find the answer to that, it would be the ultimate triumph of human reason—for then we would know the mind of God."

ILLUSTRATIONS CREDITS

COVER: David Hiser—Photographers/Aspen

FRONT MATTER: Pg. 1, Dan McCoy/Rainbow; 2-3, Gerd Ludwig; 4-5, Frieder Blickle/Bilderberg/ AURORA

TIME: THE ELUSIVE CONCEPT

Pg. 6, The Granger Collection, NY; 7, Ancient Art & Architecture Collection Ltd.; 8, Erich Lessing/Art Resource, NY; 12-13, René Milot; 15 (both), The Granger Collection, NY; 16, The Granger Collection, NY; 18-19, European Southern Observatory; 20, Erich Lessing/Art Resource, NY; 22, Visuals Unlimited

CYCLES OF NATURE

Pp. 24-25, Nicholas DeVore III—Photographers/ Aspen; 26, Adriel Heisey; 28-29, J.C. Leacock— Photographers/Aspen; 33, Eric Meola/The Image Bank; 34-35, Roger Ressmeyer/Corbis; 36-37, Slim Films; 38, Mark W. Moffett; 39, O. Louis Mazzatenta; 40, Stephen J. Mojzsis, UCLA; 41, James King-Holmes/ Science Photo Library/Photo Researchers, Inc.; 43, Slim Films; 44-45, George Steinmetz; 46, stone/Dietrich Rose; 48, Frieder Blickle/Bilderberg/AURORA; 50-51, George Grall; 52, Maggie Steber; 52-53, Cary Wolinsky; 53, Michael Yamashita; 54, Wolfgang Kaehler Photography/www.wkaehlerphoto.com; 55, Robert W. Domm/Visuals Unlimited; 56, CNRI/Phototake, NY; 57, Karen Kasmauski; 58, Dan McCoy/ Rainbow; 59, Manfred Kage/Peter Arnold, Inc.;60-61, Joel Sartore/ www.joelsartore.com; 62-63 (upper), Chris Johns, National Geographic Photographer; 62-63 (lower), stone; 64, Brooks Walker; 65, Oscar Burriel/Science Photo Library/Photo Researchers, Inc.

TIMEKEEPERS OF THE ANCIENTS

Pp. 66-67, Jim Richardson; 68, Scala/Art Resource, NY; 70 (left), stone/Robert Frerck; 70 (right), Alexander Marshack; 71, Sisse Brimberg; 72, Sisse Brimberg; 73, British Museum; 74-75, Antony Edwards/The Image Bank; 77, Stephen Trimble; 78, Erich Lessing/Art Resource, NY; 79, Guido Alberto Rossi/The Image Bank; 80-81, Bob Sacha; 82 (upper), The Granger Collection, NY; 82 (lower), Science Museum/Science & Society Picture Library; 82-83, Scala/Art Resource, NY; 84, Scala/Art Resource, NY; 85, Giraudon/Art Resource, NY; 86-87, Archivo del Stato, Siena, Italy/Roger-Viollet, Paris/The Bridgeman Art Library; 89, Science Museum/Science & Society Picture Library; 92-93, David Hiser; 94, David Alan Harvey; 95, Enrico Ferorelli; 96-97, Gerald L. French/ThePhotoFile; 98, National Maritime Museum Picture Library; 99, By Permission of the Folger Shakespeare Library; 101, The Granger Collection, NY; 102, Illustration by David Penney; 103, Science Museum/Science & Society Picture Library

MECHANIZING TIME

Pp. 104-105, Antony Edwards/The Image Bank; 106, Edwin L. Wisherd; 107, The Granger Collection, NY; 110, Guido Alberto Rossi/The Image Bank; 110-111, Erich Lessing/Art Resource, NY; 112-113, Science Museum/Science & Society Picture Library; 114, Winterthur, Uhrensammlung Kellenberger; 115, Joanna B. Pinneo; 117, Slim Films; 118-119, Michael Freeman; 121 (upper), Bruce Dale; 121 (lower), Art Resource, NY; 122 (left), Science Museum/Science & Society Picture Library; 122 (right), The Time Museum; 123 (upper left), Telegraph Colour Library/FPG International; 123 (upper right), stone/John Lamb; 123 (lower), Science Museum/Science & Society Picture Library; 125, Science Museum/Science & Society Picture Library; 126-127, Bob Sacha, by permission of The Vatican Secret Archives; 128, The Granger Collection, NY; 129, Leonia 800011-49841, courtesy National Board of Antiquities, Helsinki; 130-131 (all), The Granger Collection, NY; 134, The Granger Collection, NY; 134-135, Robert McGinnis; 137, Science Museum/ Science & Society Picture Library; 138-139, The Mapparium of the Christian Science Publishing House in Boston, photo by Bob Sacha.

FINE-TUNING TIME

Pp. 140-141, Frieder Blickle/Bilderberg/AURORA; 142, Bob Sacha; 146-147 (both), The Granger Collection, NY; 148, Harvey Lloyd/Peter Arnold, Inc.; 148-149, New York Public Library; 150, John Watney; 150-151, Bob Sacha; 153, Corbis; 154, Peter Menzel; 156 (both), The Granger Collection, NY; 157, FPG International; 159 (all), Robert F. Sisson; 160, Victor F. Kayfetz/Black Star; 160-161, FPG International; 162, Ross M. Horowitz/The Image Bank; 163, Shahn Kermani/Liaison Agency; 164, Archive Photos; 165, Stephen Elleringmann/Bilderberg/ AURORA; 166, Cary Wolinsky; 166-167, O. Louis Mazzatenta; 168, J. Pickerell/FPG International; 169, Robert Rathe/Stock Boston; 170-171, Harold & Esther Edgerton Foundation, 2000, courtesy of Palm Press, Inc.; 172-173, Alexander Tsiaras/Science Source/Photo Researchers, Inc.; 173, Bruce Dale; 175, National Institute of Standards and Technology; 176-177, Bob Sacha; 177 (upper), ESA/CE/Eurocontrol/ Science Photo Library/Photo Researchers, Inc.; 177 (lower), Bob Sacha; 179, Morten Kettel

THE TIME OF OUR LIVES

Pp. 180-181, Joel Sartore/www.joelsartore.com; 182, Jim Brandenburg; 184, Musée du Louvre, Paris/ SuperStock; 185, Steve Woit; 188-189, Joel Sartore/ www.joelsartore.com; 190-191, Lynn Johnson/ AURORA; 193, Slim Films; 194, José Azel/ AURORA; 195, stone/James Strachan; 196-197, William Albert Allard, National Geographic Photographer; 198, Sam Abell, National Geographic Photographer; 199, Michael Yamashita; 202-203,

Peter Ginter/Bilderberg/AURORA; 204, Archive Holdings, Inc./London Daily Express/The Image Bank; 205, stone/Hulton Getty; 206-207, stone/Penny Tweedie; 207 (upper), Joanna B. Pinneo; 207 (lower), Joel Sartore/ www.joelsartore.com; 208-209, Mary Jane Cardenas/The Image Bank; 210, Index Stock Imagery; 211, James King-Holmes/ICRF/Science Photo Library/Photo Researchers, Inc.; 212-213, Robb Kendrick; 215, Chip Hires/Liaison Agency

DEEP TIME

Pp. 216-217, Joe McNally; 218, NASA/Space Telescope Science Institute; 219, NASA/Science Photo Library/Photo Researchers, Inc.; 222 (upper), stone/Simon McComb; 222 (lower), adapted from David Parker/Science Photo Library/Photo Researchers, Inc.; 223, Cary Wolinsky; 224-225, Joe McNally; 229, Peter Ginter/Bilderberg/AURORA; 230-231, CERN/Science Photo Library/Photo Researchers, Inc.; 232-233, Joe Stancampiano, NGS, and Karl Luttrell, University of Michigan; 233, NASA; 234-235, Howard Koslow; 235, Bruce Dale; 236, Julian Baum/Science Photo Library/Photo Researchers, Inc.; 237, NASA/ Phototake NY; 238, Roger Ressmeyer/Corbis; 239 (upper), NASA; 239 (lower), Space Telescope Science Institute/NASA/Mark Marten/Photo Researchers, Inc.; 240-241, Art by Christopher Sloan, NGS Staff, and David Fierstein. Sources: Richard Gott, Jeff Hayes, Robert Osserman, Jeff Weeks.; 242, James A. Sugar, with Kevin Schumacher and Larry D. Kinney; 243, David A. Hardy/Science Photo Library/Photo Researchers, Inc.; 244-245, Joe McNally; 248, Peter Essick; 249, Holly Warburton.

BACK DUST JACKET: (upper) Robb Kendrick; (lower left) James King-Holmes/Science Photo Library/Photo Researchers, Inc.; (lower right) Mary Evans/Explorer Archives. Back flap: Steve Zavatski

NOTE ABOUT THE AUTHOR

JOHN LANGONE, a veteran science journalist, has served as a staff writer for *Time* and *Discover* magazines, a reporter and writer for United Press International, and science editor at the *Boston Herald*. Langone was a Kennedy Fellow in Medical Ethics at Harvard, a Fellow at the Center for Advanced Study in the Behavioral Sciences at Stanford, and a Fulbright Fellow at the University of Tokyo. A contributor to the "Science Times" section of the *New York Times* and to *Departures* magazine, the author has taught science-writing at Harvard, Boston University, and New York University. This is his 24th book. Langone lives in Old Lyme, Connecticut.

ACKNOWLEDGMENTS

The Book Division acknowledges the invaluable assistance of assistant editor Dale-Marie Herring at the inception of this book. Her dedication and clear thinking were important in conceptualizing and organizing the book's scope and content. Special thanks goes to Gillian Carol Dean for rebuilding the artwork on pages 240-241. We also appreciate the careful eye for detail that Michele Tussing Callaghan applied when reviewing the book in its final stages. For their artistry and cheerful cooperation, we offer Slim Films our sincere thanks. They created several pieces of art specially for us. To access their website, go to slimfilms.com. In addition, we extend thanks to our many colleagues at the Society who, through their cooperation and assistance, helped make *The Mystery of Time* a reality. Most particularly, we acknowledge the NATIONAL GEOGRAPHIC magazine, the Library, the Indexing Division, the National Geographic Image Collection, and the National Geographic Photographic and Imaging Laboratory.

For generously sharing their expertise and assistance, the Book Division gratefully acknowledges Bulent I.

Atalay, Mary Washington College, Fredericksburg, Virginia; Patricia Atwood, the Time Museum; Geoff Chester, U.S. Naval Observatory, Washington, D.C.; Keith Devlin, Dean, School of Science, Saint Mary's College of California; Valerie Mattingley, the National Geographic Society's United Kingdom Office; Joanne Petrie, U.S. Department of Transportation; Andrew J. Pogan; Curt Suplee; W. David Todd, Museum Specialist, Timekeeping Collection, Smithsonian National Museum of American History, Washington, D.C.; Brigitte Vinzens, Uhrensammlung Kellenberger, Winterthur, Switzerland; and the Gesell Institute of Child Development, New Haven, Connecticut.

ADDITIONAL READING

The reader may wish to consult the National Geographic Index for related articles. The following sources may also be of interest:

Barnett, Jo Ellen. *Time's Pendulum: The Quest to Capture Time—From Sundials to Atomic Clocks,* New York, Plenum Press, 1998.

Davies, Paul. *About Time: Einstein's Unfinished Revolution,* New York, Simon & Schuster, 1995.

Hawking, Stephen W. *A Brief History of Time: From the Big Bang to Black Holes,* New York, Bantam Books, 1988.

Howse, Derek. *Greenwich Time and the Discovery of Longitude,* New York, Oxford University Press, 1980.

Landes, David S. *Revolution in Time: Clocks and the Making of the Modern World,* Cambridge, Mass., Harvard University Press, 1983.

Composition for this book by the National Geographic Society Book Division. Printed and bound by R. R. Donnelley & Sons, Willard, Ohio. Color separations by Quad/Graphics, Martinsburg, West Virginia. Dust jacket printed by Miken Inc., Cheektowaga, New York.

INDEX

Boldface indicates illustration.

THE MYSTERY OF **TIME**
By John Langone

Published by the National Geographic Society

John M. Fahey, Jr., *President and Chief Executive Officer*

Gilbert M. Grosvenor, *Chairman of the Board*

Nina D. Hoffman, *Senior Vice President*

Prepared by the Book Division

William R. Gray, *Vice President and Director*

Charles Kogod, *Assistant Director*

Barbara A. Payne, *Editorial Director and Managing Editor*

Staff for This Book

Martha Crawford Christian, *Editor*

K. M. Kostyal, *Text Editor*

Greta Arnold, *Illustrations Editor*

Suez Kehl Corrado, *Art Director*

Mimi Harrison, *Illustrations Researcher*

Elisabeth B. Booz, *Senior Researcher*

Marilynh "Minh" Le, *Researcher*

R. Gary Colbert, *Production Director*

Richard S. Wain, *Production Project Manager*

Lewis R. Bassford, *Production*

Meredith C. Wilcox, *Illustrations Assistant*

Peggy Candore, *Assistant to the Director*

Robert W. Witt, *Staff Assistant*

Elisabeth MacRae-Bobynskyj, *Indexer*

Manufacturing and Quality Control

George V. White, *Director*

John T. Dunn, *Associate Director*

Vincent P. Ryan, *Manager*

Phillip L. Schlosser, *Financial Analyst*

Library of Congress Cataloging-in-Publication Data

Langone, John, 1929-
   The mystery of time : humanity's quest for order and measure / John Langone.
      p.   cm.
   Includes bibliographical references and index.
   ISBN 0-7922-7910-7 (regular) -- ISBN 0-7922-7911-5 (deluxe)
         1. Time--History. I. Title.

BD638 .L26 2000
115--dc21   00-041812

The world's largest nonprofit scientific and educational organization, the National Geographic Society was founded in 1888 "for the increase and diffusion of geographic knowledge." Since then it has supported scientific exploration and spread information to its more than nine million members worldwide.

The National Geographic Society educates and inspires millions every day through magazines, books, television programs, videos, maps and atlases, research grants, the National Geography Bee, teacher workshops, and innovative classroom materials.

The Society is supported through membership dues and income from the sale of its educational products. Members receive NATIONAL GEOGRAPHIC magazine—the Society's official journal—discounts on Society products, and other benefits.

For more information about the National Geographic Society and its educational programs and publications, please call 1-800-NGS-LINE (647-5463), or write to the following address:

National Geographic Society
1145 17th Street N.W.
Washington, D.C. 20036-4688 U.S.A.

Visit the Society's Web site at
www.nationalgeographic.com.